Finding Birds in the National Capital Area

Finding Birds in the National Capital Area

Second Edition

Claudia Wilds

Smithsonian Institution Press
Washington and London

Editor and typesetter: Peter Strupp/Princeton Editorial Associates
Production editor: Jack Kirshbaum
Designer: Alan Carter
Illustrator: Doreen Curtin
Cartographers: James D. Ingram and Brian T. Davis
Cartographic Services Laboratory, University of Georgia, Athens.

Library of Congress Cataloging-in-Publication Data

Wilds, Claudia P.
    Finding birds in the national capital area / Claudia Wilds.—2nd ed.
        p.   cm.
    Includes bibliographical references and index.
    ISBN 1-56098-175-X
        1. Bird watching—Washington Region.   2. Birds—Washington Region.
    3. Bird watching—Virginia.   4. Birds—Virginia.   5. Bird watching—Maryland.
    6. Birds—Maryland.   I. Title.
    QL683.W37W54   1992
    598′.0723475—dc20   92-7026
                                                                                                    CIP

British Library Cataloging-in-Publication data available

06   05   04   03   02   01   00   5   4   3

Cover illustration: Pileated Woodpecker

Frontispiece: Peregrine Falcon

For
Ives and Evan
Lola and Ted
whose friendship is reward enough
for having taken up birding

Contents

THE EASTERN SHORE

SOUTH OF CHESAPEAKE BAY

SPECIAL PURSUITS

APPENDIXES

Maps of the **National Capital Area** and the **District of Columbia and Arlington County** can be found at the back of this book.

Preface to the Second Edition

The first edition of this book began to go out of date when it was still in press—the main entrance to Huntley Meadows was shifted to the other side of the park. In the years since then an overgrown nursery famous for its wintering owls became an industrial park, whole sections of Washington became unsafe for birding, and a favorite turf farm was closed to visitors because a few birders would not keep off the grass. Common and widespread birds like Northern Bobwhites and Eastern Meadowlarks became uncommon and local; uncommon and local species like Red-cockaded Woodpeckers, Bewick's Wrens, and Henslow's Sparrows virtually vanished.

On the other hand, Huntley Meadows evolved into one of the finest wetlands in the region, new sites for hawk watching proliferated, and the gull-watching mecca of Conowingo Dam was discovered. The cleanup of the Potomac combined with the right amount of rainfall to produce hydrilla mats on the river and several years of spectacular waterfowl numbers. Bald Eagles and Eastern Bluebirds became easy to find, Lesser Black-backed Gulls could be seen at every reservoir and landfill, and Brown Pelicans exploded up the coast.

No bird-finding guide can be completely trusted (quite aside from its variable accuracy in pinpointing locations of particular species), given the rapid rate of change. This edition contains an even higher ratio of public lands to private lands than the first, but parks close, or they change hours and fees and roads. Mileages are almost always a bit inaccurate, especially those measured from highways with interchanges instead of crossroads. Many of the region's roads are being widened or rerouted, and future intersections and points of access cannot be foreseen. Delaware and Maryland do not always put up road numbers or names at intersections; Virginia is good about road numbers but may assign a different name from the one on your map. (Still, with so much water and so many highways setting limits, you cannot stay lost for long.)

Although this guide will take you to many of the best-known and best-loved birding areas in the region, there are still corners of every state and even of the District of Columbia that have hardly been investigated at all. If racking up a big life list (or state list or annual list) is not your only reason for birding, try instead or in addition the pleasures of locality birding or working a patch: define an area that you can bird thoroughly many times a year, and make yourself the most knowledgeable person in the world about the birds of that area.

To learn what habitats each species uses for nesting or feeding, when they come and go, how changes of climate or habitat affect populations, which are the regular visitors and which the vagrants, is as delightful an experience as keeping an annual state list and far more instructive. Moreover, as an expert on the birds in your patch, with all your sightings of every species in it meticulously recorded by date and number of individuals, you become a resource that may suddenly be in demand when a new road or development threatens the area.

The Maryland Ornithological Society combines a concession to the existence of listing fever with encouragement of thorough exploration of the state. It promotes the keeping of county lists, the totals of which are published annually in the *Yellowthroat*. Seeing the largest possible number of species in a county that lacks several major habitats (high-altitude forests, for example, or salt water) requires many days in the field and close attention to the landscape. If you have a distaste for competitive listing or do not bird in Maryland, the game can be played privately anywhere, especially in the neglected counties of Virginia that are laced with untraveled back roads. If a whole county is too large, choose a part of one. Finding birds is an intense joy; finding new places to see birds, especially disappearing or local and uncommon ones, is a joy and a contribution to knowledge, even if you do not publicize your discoveries.

In addition there are a dozen different ways (listed in Appendix C) in which you can engage in cooperative birding activities, collecting data that need to be gathered for environmental studies, for protection or restoration of declining species, or simply for increased understanding of the avian world.

Acknowledgments for the
Second Edition

This edition, like the first, owes not just its accuracy but its very completion to an army of birders. Not including the dozens who simply answered a question or two or pointed out an error in the first edition or a change in a telephone number, I am indebted to many friends, strangers, and strangers who became friends in the course of preparing this manuscript.

Foremost among them is David Czaplak, who greatly expanded and completely redrafted the chapter on the District of Columbia, describing and showing me places I never had known about, deleting unsafe and unproductive sites, and making the text far more authoritative than before.

George Jett and Armand Miller, in response to a very general request for help, collected detailed information on parks and refuges in Maryland and Virginia, respectively, and sent me maps, bird lists, and commentary that were invaluable in making decisions on sites that should be included and what should be said about them.

Gene Hess reviewed and improved the chapters on the Delaware refuges and northern Delaware, prepared a list of Delaware rarities at my request, and showed me the pleasures of White Clay Creek, a prize addition to the book.

Ed Weigel spontaneously shared his unique expertise on Huntley Meadows, as did Erika Wilson on the Mason Neck parks.

Several birders ran routes from the first edition in order to test their accuracy, in some cases under strenuous and trying conditions, and enriched my files with maps, bird lists, corrections, and new ideas. They include Jim and Beth Bullard, Robert Caswell, Jim Gruber, Robert Hilton, Michael O'Brien, Gene Scarpulla, Paul and Ann Smith, Richard Szabo, Mary Ann Todd, Virginia and Allen Valpey, and Frank Witebsky.

Rick Blom, Gene Hess, Teta Kain, Michael O'Brien, Bob Ringler, and Erika Wilson carefully reviewed the "Species to Look For" chapter, supplying important information about the current status and distribution of species in the region and making valuable suggestions on old sites to delete or modify and new areas to include.

Paul Baicich, Roger Clapp, Bettye Fields, William Frech, Lola Oberman, Teresa Simons, Don Stein, Len Tember, Hal Wierenga, and Erika Wilson guided me around favorite areas and gave me unique perspectives and knowledge. Bob Behrstock took notes for me on another tour, and my brother Lea served as a patient chauffeur.

John Bazuin, Larry Bonham, Daphne Gemmill, David Mehlman, and Sally Ann Waldschmidt volunteered a variety of exceptionally useful information.

In response to my requests, Lynn Badger, John Bjerke, Michael Boatwright, John Fussell, Mark Garland, Delores Grant, David Heskell, Greg Kearns, Bob Keedy, Harry LeGrand, Greg Lewis, Pat Moore, Bill Murphy, Teresa Simons, Keith Van Ness, and Erika Wilson all reviewed new or old text and improved its accuracy and clarity and added new information. Carol Beyna, Mike Boatwright, Paul Dumont, Carl Garner, Cameron Lewis, Paul Nistico, and Brian Patteson responded promptly and concretely with answers to my inquiries. Susan Cook made the task of revision infinitely easier by converting the first edition into disk files for my computer.

Preface to the First Edition

This book is the outgrowth of a series of articles prepared for the *Audubon Naturalist News,* the newsletter of the Audubon Naturalist Society of the Central Atlantic States. The society is an independent, regional natural history and conservation group based in Washington, D.C., and 95 percent of its members live in the National Capital Area.

The society's strong program of field trips has been, in recent years, the best introduction available to the birding sites of the region, and many of the locations described here are on its calendar of regularly scheduled bird walks.

These and several other areas were brought to my attention as editor of the Voice of the Naturalist, also sponsored by the society. This weekly telephone recording of news for birders features the locations of rare or hard-to-find species sighted in the region, reports of migration and invasions of irruptive species, and sites that local birders are finding especially productive.

A number of chapters are the direct result of asking a regular contributor to the Voice for a tour of a favorite locale. A few sites are included because a listener to the Voice or a reader of the *News* or a friend feared that I was overlooking a place that he or she wanted to share.

Regardless of source, the selections are a reflection of my own knowledge and prejudices. There may be superb birding sites in Virginia southwest of Washington or in Maryland northeast of Washington or along the lower Potomac, but I do not know about them. The very occasional sightings reported from these areas are almost always of species that are just as easy to see elsewhere close to the capital. By contrast, although I like to think that the reader has an abundance of land-bird sites to choose from, my own preference for water birds and for coastal habitats, especially my affection for Chincoteague, is hard to miss.

No attempt has been made to treat with evenhanded thoroughness all the worthwhile locations in the area. Good birding spots in and around other towns and cities are included only if they are of particular interest or so fruitful that they merit a special outing or stopover on travels from the District of Columbia. Several excellent publications, listed in Appendix B, cover other sites both within and beyond the range of this book. In particular, Hawk Mountain in Pennsylvania and Cape May in New Jersey, though visited regularly by Washington area birders, are so thoroughly described elsewhere that I could not bring myself to repeat readily available information.

This may be as good a point as any to mention the mileage problem. Almost every route was checked at least once, and some were run in three different vehicles. Even the shortest distances may be 0.1 mile off from your odometer; on longer stretches a 0.1 mile discrepancy for every 3 miles or so appears to be entirely normal. Wherever possible, road names or numbers or other landmarks have been included to minimize confusion, and the maps may do more than prose can to get you where you want to go.

Acknowledgments for the First Edition

When I began looking at birds in 1970, one of my most joyful discoveries was a delightful, lucid, and informative series of articles collectively called "Where to Go," written by Carl W. Carlson and published in the *Atlantic Naturalist* in the 1960s. They guided me flawlessly through dozens of the finest birding areas near Washington, giving me continuing pleasure for several years. Gradually, interstate highways and condominiums paved over some of the nesting sites and migrant traps Carl had described. Only his resistance to updating and compiling his fine series made the present book a reasonable undertaking. Though none of his articles was used as a direct source, they must all be acknowledged as the basic course on finding birds around Washington that inspired this effort, and as a model impossible to emulate.

For many of the sites described herein I am heavily indebted to local birders, who generously shared their knowledge of specific areas, some spending an entire day or several of them supplying not only a deluge of information but also the greater gift of companionship. My fervent thanks go to Jackson Abbott (Alexandria), Roger Anderson (Long Branch), Harry Armistead (Blackwater, Hooper Island, Elliott Island, Deal Island), John Bazuin (Loudoun County, Lake Anna), Rick Blom (Baltimore Harbor, Baltimore County, Fort Smallwood, St. Mary's County), Dan Boone (Washington County), Terry Boykie (Long Branch), Dave Czaplak (District of Columbia), Owen Fang (Alexandria), Tony Futcher (District of Columbia), Delores Grant (Meadowside), John Gregoire (Jug Bay), Evan Hannay (the Pulpit, geographic research), Kerrle Kirkpatrick (Linden Fire Tower), Valerie Kitchens (Long Branch), Floyd Parks (Kent County), Bob Patterson (Jug Bay, Merkle), Bob Ringler (Baltimore Harbor), Stan Shetler (Algonkian Park), Jim Stasz (map for Back River Sewage Treatment Plant), Len Teuber (Myrtle Grove, Oxon Hill Children's Farm, Highland County, Shenandoah National Park), Bill Thomas (Long Branch), Barry Truitt (Virginia Coast Reserve), Craig Tufts (Claude Moore Center), Keith Van Ness (Seneca Creek), Dean Weber (Garrett County), and Hal Wierenga (Sandy Point, Deal Island, owl-finding).

I am especially grateful to Joy Aso and John Malcolm, two volunteers who divided the tedious job of checking the mileages and directions for most of the trips in this book. Between them they drove nearly 3,000 miles, and then transcribed notes, sketched maps, and drafted improved directions. I also thank Sharon Malcolm for navigating for John on some of his trips and for being so patient while he went off to check other routes.

Finally, I wish every author had good fortune equal to mine in having an ally like Eve Bloom, my typist, counselor, and friend, who saw to it that a thousand ambiguities and inconsistencies bit the dust.

The Region
and Its Birds

Introduction   1

A third of the species of birds that inhabit or regularly visit the continental United States and Canada may be found in the course of a year within the Capital Beltway. More than 350 species may be seen within a four-hour drive of the District of Columbia.

The location of the national capital on the line between piedmont and coastal plain, less than three hours away from both 4,000-foot mountains and ocean beaches, offers bird watchers of the Washington area an exceptional variety of habitats in which to study birds. Ten major National Wildlife Refuges and innumerable national, state, and regional parks and wildlife areas provide assurance of permanent access to important sanctuaries throughout the region.

Nearly 200 species regularly breed in the region, ranging from Saw-whet Owls, Hermit Thrushes, and Nashville Warblers in far western Maryland to Wilson's Plovers, Sandwich Terns, and Swainson's Warblers in southeastern Virginia. More than half of them breed in the counties around the beltway, and Washington itself offers nesting Fish Crows, Veeries, Prothonotary Warblers, and Louisiana Waterthrushes, among other species.

Migration goes on virtually nonstop, reaching ebb tide only from mid-December to mid-February and from mid-June to early July. A remarkable number of major migration routes converge on the region or pass through it: the mountain ridges, the Potomac and the Susquehanna rivers, both shores of the Chesapeake Bay, and, of course, the Atlantic coastline.

Raptors, especially buteos, use the ridges mainly in fall, passerines in both fall and spring. Waterfowl and gulls are the most conspicuous migrants on the rivers in fall and early spring, but in May one can see thousands of nighthawks and swallows moving steadily upstream, and riverside trails are alive with transient passerines. Loons, waterfowl, and raptors stream up or down the shores of the Chesapeake and are often easy to see where the width of the bay is constricted. The coast is incomparable for its abundance of migrating loons, gannets, cormorants, herons and ibis, Snow Geese, Brant, sea ducks, falcons, shorebirds, and fall swallows and other passerines. Aside from these principal flyways, migrant passerines seem to blanket the area, appearing in every faintly appropriate habitat, most abundantly close to streams and marshes.

The importance of the region as a wintering ground lies overwhelmingly in the habitats of the coastal plain, attested to by the concentration of wildlife refuges and the high species totals on the

Christmas Bird Counts, especially on southern Delmarva. (The Cape Charles count averages more than 160 species.) High numbers of waterfowl—swans, geese, sea ducks, and diving ducks in particular—are to be found mainly on the Eastern Shore but, for many species, in the sheltered waters from Baltimore and Washington to Point Lookout as well. Raptors are plentiful both on the coastal plain and in the piedmont. Huge flocks of gulls gather on beaches, breakwaters, sandbars, landfills, sewage treatment plants, farm fields, and golf courses all around the coastal plain and at selected sites in the piedmont near the Fall Line and open water. Hundreds of thousands of crows, blackbirds, grackles, cowbirds, and starlings are distributed throughout the lowlands of the region. Among the wintering finches and sparrows abundant everywhere but on the Appalachian Plateau, the invading House Finch is now as common as the American Goldfinch. The unpredictable arrival of irruptive species like Evening Grosbeak, Pine Siskin, and Common Redpoll lends anticipatory excitement to winter land birding.

Thus for resident and visiting birders alike, the National Capital Area offers fine opportunities year-round. The pages that follow attempt to provide useful information about the landscape, habitats, and weather that one can expect in the region and a species-by-species sketch of seasonal and geographic abundance, preferred habitats, and likely locations.

The heart of the book is the section on birding sites. Those closest to Washington are grouped geographically, but most birders are likely to sample no more than two or three in a day. Beyond the suburban counties, suggested tour routes—most of them measured from the Capital Beltway—describe excursions suitable for a day or a weekend or, in combination, several consecutive days of good birding.

The appendixes contain information about natural history and ornithological societies, maps and other publications, activities for birders in the National Capital Area, procedures for reporting rarities to local records committees, and ethics for birders.

2 Geography and Climate

THE REGION DEFINED

The region covered by this book has a configuration like a gerrymandered congressional district, reasonable for birding out of Washington, but perhaps bizarre-looking to readers from the rest of the country. It includes sites in all parts of Maryland and Delaware, the Virginia piedmont and mountains north of an east-west line just below Charlottesville, and the Virginia coastal plain close to Washington, south of the James River and east of the Chesapeake Bay.

The North Carolina Outer Banks are beyond the National Capital Area by any rational definition, but birders do go there from Washington and published information about the way to bird the Banks is all too scanty. Though included in the book, they should be viewed as in an adjacent region with a somewhat different avifauna.

PHYSIOGRAPHY AND HABITATS

The region is divided into five well-defined and traditional physiographic provinces, two of which have been lumped together here for ornithological simplicity. Those at the western and eastern extremes have the most distinctive habitats and bird life, but each shares many characteristics with its neighbors.

The Appalachian Plateau. This highland province, covering almost all of West Virginia, reaches its eastern outpost in Maryland's Allegany County along the ridge of Dan's Mountain. On the Maryland plateau the valley floors do not sink below 1,500 feet; in Garrett County they average 2,500 feet, and the ridges and peaks that rise above them are modest. Maryland's highest mountain, with its summit in the extreme southwest corner of the state, is Backbone Mountain, only 3,360 feet high. The hills and valleys here are somewhat randomly oriented, not strongly parallel as they are farther east.

The mountains, except for their rocky summits and sheer cliffs, are typically covered with second-growth oak, maple, and pine, though some of the north-facing slopes have stands of hemlock. The wide valleys, known locally as "glades," are mostly bluegrass pasture land, though many are under cultivation. Along the streams, notably at Swallow Falls, one can find groves of hemlock, spruce, and pine. Deep Creek Lake provides a major man-made stillwater habitat for migrant water birds.

Unique for the region are the boreal bogs of heath and sedge-meadow, broken by scattered rhododendron and alder bushes and clumps of spruce, hemlock, birch, and maple; the most famous is Cranesville Swamp on the border with West Virginia.

Dividing the plateau from the next province to the east is the Allegheny Front, an escarpment on a virtually straight northeast-southwest line from Cumberland, Maryland, to Bluefield, Virginia, entirely to the west of northern Virginia.

Ridge and Valley (including the physiographically distinct Blue Ridge). The outstanding characteristic of this province is the tightly folded landscape shaped into a series of long ridges and valleys in parallel lines, all running northeast-southwest. In the west the valleys are narrow, but east of Clear Spring in Maryland the broad Hagerstown Valley marks the northward extension of the Valley of Virginia, up which the two forks of the Shenandoah meander to their junction at Front Royal. To the east of this valley the Blue Ridge marks the eastern boundary of the province in Virginia. In Maryland the lower counterparts of the Blue Ridge are South Mountain and Catoctin Mountain, with the Middletown Valley in between.

The entire province is much higher in Virginia than in Maryland, with most ridges above 3,000 feet and some peaks over 4,000 feet. Harrisonburg in the Shenandoah Valley lies at over 1,900 feet and Monterey in a valley west of Jack Mountain is exactly 3,100 feet above sea level.

In Maryland, on the other hand, the altitude of the valleys is no higher than most of the piedmont, at 500–600 feet, and the highest ridge, just east of Cumberland, is Warrior Mountain at only 2,135 feet.

The broader valleys are heavily farmed; the valley floor is under cultivation and the slopes are rich in apple orchards. The narrower western valleys are usually bluegrass pastures or hay fields, with boggy marshes near the headwaters of highland creeks; relatively little land is still under the plow, and abandoned fields are easy to find.

The habitats of the Blue Ridge and the western Virginia ridges are quite similar to those of the mountains of the Appalachian Plateau. The northern conifers are less common, however, and the bogs are missing. Even the lower highlands in Maryland, rising steeply from the flood plain, provide the altitude and climate that attract species to the oak-beech-hickory forests that are rare in or absent from similar habitats to the east.

The bottomland woods along the Potomac in this province and the next, all the way from Cumberland to Washington, are traversed along the Maryland side by the old Chesapeake and Ohio (C&O) Canal and the highly accessible towpath beside it.

The Piedmont. The national capital lies on the edge of this gently rolling plateau province. The western limit runs along the foot of the Catoctin Mountains in Maryland, jogs west to Harpers Ferry, and continues along the base of the Blue Ridge. In the east the piedmont ends at the Fall Line, where the soil changes from clay to sand and the rivers tumble through rocky rapids to the navigable tidal waters of the coastal plain. Interstate 95, curving south through Wilmington, Baltimore, Washington, and Richmond, runs just above the Fall Line from the Delaware River to the North Carolina border.

The piedmont is divided by two great rivers, the Susquehanna and the Potomac (and its tributaries), and by four other important streams, the Gunpowder, the Patapsco, the Patuxent, and the Rappahannock. Virtually all the still waters of the province are in man-made reservoirs, recreational lakes, and farm and fish ponds. All the rivers—especially the Potomac and the Monocacy, which flows into it—are bordered by outstanding examples of flood plain forest, dominated by giant sycamores, maples, elms, and hackberry. Georgetown Reservoir in Washington and Frank and Clopper lakes in Montgomery County, Maryland, are among the closest lakes to the capital. Piedmont marshes of exceptional interest are at Hughes Hollow and Lilypons.

Where it has not been paved over, the province, with an altitude that ranges from 300 to about 800 feet, consists of active and abandoned farmland and of forest and woodlots in various stages of succession. Cropland, hay fields, and pastures and turf farms, divided and edged by bird-rich hedgerows, line the back roads of nearby Loudoun County in Virginia and western Montgomery County and southern Frederick County in Maryland.

Recently abandoned farmland is marked by fields of broomsedge and goldenrod, dotted with red cedar and saplings of red maple and other hardwoods. The red cedar and, more locally, Virginia pines grow into solid groves. In later years deciduous trees invade the shaded understory, then take over: oak, beech, hickory, and tulip tree, with red maple, dogwood, redbud, and sassafras among the trees in the understory.

The forested areas increase as one heads south in Virginia and the proportion of pines to hardwoods expands as well.

Coastal Plain. Slashed into three distinctive sections by Chesapeake Bay and the James River, the coastal plain is marked by the narrowest range of altitude (0–300 feet) and the greatest diversity of habitats in the region.

The Western Shore (a term normally applied only to Maryland west of Chesapeake Bay) shares almost all of its characteristics with the Virginia shoreline from Arlington to Hampton. The somewhat hilly landscape near the Fall Line flattens out in the south, close to the bay. Upland oak-hickory forest is typically found inland; swampy bottomland woods and pine stands (mostly of Virginia pine) are extensive. The predominant farmland is in row crops, including tobacco, with relatively little pasture or hayfield habitat. Fresh to brackish marshes are scattered throughout; the finest wild rice marshes in the east line the Patuxent near Jug Bay, but Baltimore County, Dyke Marsh, Huntley Meadows, Sandy Point, and Point Lookout all offer excellent habitats for rails, marsh wrens, and herons. Wooded swamps range from Mason Neck to Myrtle Grove to an isolated bald cypress swamp in Calvert County. The convoluted shoreline of Baltimore County, along with urban environments like landfills and sewage plants, attracts gulls and waterfowl in great numbers.

The nearly level Eastern Shore (a term interchangeable with *Delmarva Peninsula*) is under cultivation from end to end wherever there is sufficient drainage and the land is unprotected.

Among the interspersed patches of bottomland forest two distinctive habitats draw avian specialties. The stands of loblolly pine increase in abundance southward from their northern limit between Kent Island (at the east end of the Chesapeake Bay Bridge) and Rehoboth, Delaware; they are readily accessible at Blackwater and Chincoteague refuges. Bald cypress, loblolly pine, water oak, and red bay mark the extensive and famous swamp along the Pocomoke River.

The superb marshes of Delmarva range from the huge freshwater marshes on either side of the Chesapeake and Delaware Canal to the tidal marshes that line Delaware Bay and both sides of the peninsula from Rehoboth and Cambridge south.

Off the shore of the lower bay numerous wooded islands, marsh islands (locally called "tumps"), and sandbars are the nesting sites of many of the region's colonial water birds. The shoreline itself is so dominated by long, twisting creeks, bays, coves, and inlets that it is one of the nation's greatest concentration points for wintering waterfowl.

The ocean beaches from Cape Henlopen to Ocean City are primarily a human playground, enhanced for winter birders by two jetty-lined inlets that provide (along with the Chesapeake Bay Bridge-Tunnel at the tip of Delmarva) the region's principal rocky coast habitat.

The coastal islands from Assateague south, with their miles of beaches, dunes, salt marshes, tidal flats (and, on some, pine woods), are not only major water-bird nesting areas but also a crucial stopover for migrant shorebirds and a highway for southbound raptors.

Finally, the impoundments of the federal and state wildlife refuges, found the length of Delmarva, are important to tens of thousands of birds all year, populations that change markedly with the season and the variations in water level.

Southside Virginia, that part of the coastal plain south of the James River and Chesapeake Bay, is conspicuously more southern than the rest of the region, warmer and wetter, with regular avian residents and visitors that are rare to accidental farther north. The remaining undeveloped land east of the Elizabeth River is similar to the swampiest, marshiest stretches of Delmarva. To the west the Great Dismal Swamp, intersected by dykes along canals, is a vast and beautiful southern swamp of bald cypress, tupelo, and white cedar. The sandy farmland south of the James is largely devoted to peanut growing, but the feature of greatest ornithological importance is the broken and disappearing loblolly pine forest that is home to most of Virginia's small and endangered population of Red-cockaded Woodpeckers.

CLIMATE

The Washington climate is both temperate and humid, a condition that makes for a green, lush environment, but one in which the cold is dank and penetrating and the heat sultry and oppressive. The range of temperature in any one year is typically 0°–100°F, except in the highlands, where it may drop to –30°. Near the coast and Chesapeake Bay the thermometer rises above 90° an average of twenty days a year; in the piedmont and the valleys and lower ridges, around thirty-five days; and on the Appalachian plateau and the high western ridges, ten days or fewer. In winter the coastal temperatures drop below freezing fifty days (south) to one hundred days (north) a year; on the plateau, about one hundred and sixty days.

If the extremes are painful, the norm is tolerable: an average July day in Washington ranges from 69° to 87° and an average January day ranges from 29° to 44°. Spring and fall are usually long and pleasant, particularly April, May, October, and November, but temperatures can fluctuate sharply.

There are some one hundred days a year with measurable precipitation, totaling 35–47 inches. Norfolk averages ten days with snow cover; Washington, about twenty; Garrett County, about seventy. Rain is more common in summer when thundery tropical air masses move in from the southwest; but in one year in three the summer weather is dominated by "Bermuda highs," when a high-pressure system stays immobile just off the coast and severe drought afflicts the region, especially the coastal plain.

The humidity is almost always high, from a low average of 60–65 percent in late winter and early spring to a high average of 75–80 percent in late summer and early fall. At dawn on an August morning on the coast it can easily be higher than 95 percent.

Species to Look For   3

The region treated in this book covers not only diverse geographical provinces but a substantial range of latitude and climate as well. The abundance of any one species may vary as much from east to west and from south to north as it does from month to month, and any precision on arrival and departure dates would have to be based on a much smaller territory. The species accounts that follow are based primarily on the Maryland and Virginia checklists and on bird lists from a dozen refuges and parks, including Bombay Hook. *Birds of*

Delaware is still in press at this writing, and the North Carolina Outer Banks are too far south to share all the patterns of the Delmarva coast. (The Outer Banks should not, as mentioned earlier, be considered as lying within the National Capital Area, but rather as the focus of an extraterritorial expedition.)

These descriptions of the species that are found with some regularity only suggest the likely provinces, habitats, and months in which to look for each of them. The recommended locations, except where a species is listed as locally common or uncommon, are by no means the only ones in which to look. A species listed for Loudoun County, for example, is as likely to be found in southern Frederick County; birds found along the C&O Canal can be expected in parks across the Potomac River. For species that are common and widespread on the coastal plain and in the piedmont, only a few sites close to Washington are mentioned; some birds are so universal that any park with appropriate habitat will do.

Most birds that are summer residents of the mountain counties are found during another season much closer to the nation's capital. Many Eastern Shore species have a vastly more restricted range, and some are virtually confined to the coast itself. Birds listed in "Delmarva refuges" are as likely to be in Blackwater as in Chincoteague; birds in "coastal refuges" are much more reliable within a mile or so of Delaware Bay and the Atlantic Ocean. Marsh nesters are especially selective and are often absent from entire areas that look no different from the ones where they are reliably present. In this book *Little Creek* refers to Little Creek Wildlife Management Area in Delaware, not to Little Creek, Virginia.

The terms *common, uncommon,* and *rare* are necessarily vague, partly because of the size of the subregion to which any of them may be applied, partly because the published sources use the terms quite differently. In general, it pays to pick up a local bird list in any refuge or park that has one and to seek out the birders (if any) on the staff for exact information on particular species.

Eager listers may notice a shortage of exotic species in the pages that follow, including most Old World waterfowl except Eurasian Wigeon. Barnacle, Lesser White-fronted, and Egyptian geese; Common and Ruddy shelducks; Garganeys; and Tufted Ducks are among the many reported waterfowl that the keepers of the state checklists regard with cautious skepticism, along with Cinnamon Teal, in view of the numerous collectors and breeders of ornamental waterfowl in the area.

Sightings of all species marked by an asterisk (*) in the main list below as well as sightings of all species on the Rarities and Accidentals list at the end of this chapter should be documented, and the documentation should be submitted to the state records committee, with a copy to the appropriate regional editor of *American Birds.* (See Appendix D for the appropriate procedure.)

The listing of species below follows the standard taxonomic sequence of the American Ornithologists' Union (AOU) *Check-list of North American Birds,* 6th edition (1983), and the 35th, 36th, 37th, and 38th supplements. This is the sequence generally followed by the *National Geographic Society Guide to North American Birds.*

Red-throated Loon, *Gavia stellata:* Common from October to May in Virginia coastal waters; less common in lower Chesapeake Bay. More common offshore, less common in sheltered waters than Common Loon. Common migrant in Delaware and Maryland in November, April, and May. Cape Henlopen, Ocean City, Chincoteague.

Common Loon, *Gavia immer:* Common in Virginia from October to May, chiefly in coastal waters and lower Chesapeake Bay; in migration also common in Maryland and Delaware waters, including inland lakes and rivers (November, April, and May). A few oversummer. Violette's Lock in migration; Cape Henlopen, Ocean City, Chincoteague, Fort Monroe.

Pied-billed Grebe, *Podilymbus podiceps:* Common from September to April, rare in summer, chiefly in sheltered fresh and brackish waters throughout region. Tidal Basin, Georgetown Reservoir, Huntley Meadows, Deal Island.

Horned Grebe, *Podiceps auritus:* Fairly common, but decreasing, from October to May, chiefly in Virginia coastal waters, lower Chesapeake Bay, and major Virginia rivers. Violette's Lock in migration; Cape Henlopen, Ocean City, Chincoteague, Point Lookout.

Red-necked Grebe, *Podiceps grisegena:* Rare from November to April, most often seen in March in Chesapeake Bay, and tidal rivers and the larger lakes and reservoirs. Baltimore Harbor, Potomac River from Wilson Bridge to Mount Vernon, Black Hill Park.

Eared Grebe,* *Podiceps nigricollis:* Rare visitor, usually near Virginia coast in fall and winter. Increasingly regular in Chesapeake Bay and lakes in Maryland piedmont in spring migration. Recorded from August to April. Craney Island, Chincoteague; Chesapeake Beach and North Beach in spring.

Northern Fulmar, *Fulmarus glacialis:* Rare and irregular well offshore, chiefly from December to March.

Cory's Shearwater, *Calonectris diomedea:* Uncommon well offshore from late May to July and in October, fairly common in August and September.

Greater Shearwater, *Puffinus gravis:* Fairly common in May and June, uncommon in summer and fall, well offshore. Occasionally seen from shore, especially after storms.

Sooty Shearwater, *Puffinus griseus:* Fairly common well offshore in late May and early June, rare in summer. Occasionally seen from shore.

Manx Shearwater, *Puffinus puffinus:* Rare but regular well offshore, with sightings in all seasons, especially in winter.

Audubon's Shearwater, *Puffinus lherminieri:* Uncommon well offshore from June to September; more common off North Carolina.

Wilson's Storm-Petrel, *Oceanites oceanicus:* Common offshore, uncommon in Delaware Bay and lower Chesapeake Bay, from May to September.

Leach's Storm-Petrel, *Oceanodroma leucorhoa:* Rare offshore from mid-May to mid-October.

Northern Gannet, *Morus bassanus:* Common offshore from late October to April; abundant November, March, and April; uncommon in May. Frequently seen along coast from shore and around Chesapeake Bay Bridge-Tunnel, and in Chesapeake Bay during spring migration.

American White Pelican,* *Pelecanus erythrorhynchos:* Rare but regular winter visitor at Chincoteague, rare all year along Virginia coast and in lower Chesapeake Bay.

Brown Pelican, *Pelecanus occidentalis:* Uncommon but increasing in coastal Delaware and Maryland from May to October, fairly common in coastal Virginia from April to December; uncommon in lower Chesapeake Bay.

Great Cormorant, *Phalacrocorax carbo:* Uncommon but increasing in recent years, from October to May. Usually on breakwaters and channel markers near coast and in Chesapeake Bay. Regular at Cape Henlopen, Indian River Inlet, Ocean City, Chesapeake Bay Bridge-Tunnel.

Double-crested Cormorant, *Phalacrocorax auritus:* Abundant in migration, chiefly from March to May and from August to November, uncommon to rare throughout the year, mostly along the coast, especially in southeast Virginia. Shearness Pool at Bombay Hook, Ocean City Inlet, Chincoteague, Hunting Bay.

Anhinga,* *Anhinga anhinga:* Accidental on coastal plain, usually seen flying overhead. Recently a frequent visitor to Stumpy Lake.

American Bittern, *Botaurus lentiginosus:* Uncommon in migration, from late March to late April and from mid-September to late October, and in winter on Delmarva. Rare local resident. In fresh and brackish marshes. Huntley Meadows, Deal and Elliott Islands.

Least Bittern, *Ixobrychus exilis:* Uncommon from May to September in fresh and brackish marshes on coastal plain; rare and local elsewhere. Huntley Meadows, Dyke Marsh, Lilypons, Deal Island.

Great Blue Heron, *Ardea herodias:* Common all year on coastal plain, uncommon elsewhere. Marshes, vegetated shores of ponds, lakes, streams. Typically solitary except near breeding colonies. Hunting Bay, Pennyfield Lock, Myrtle Grove, Deal Island, all refuges.

Great Egret, *Casmerodius albus:* Common and widespread from April to October in Delmarva marshes and southeast Virginia refuges, and along lower Potomac; uncommon to rare in winter. Rare spring visitor inland, uncommon in late summer and fall. All coastal refuges, Baltimore Harbor, Hunting Bay, Point Lookout.

Snowy Egret, *Egretta thula:* Abundant from April to September in Delmarva marshes, uncommon to rare in winter. Rare in spring, uncommon after breeding inland. All refuges on Delmarva and marshes along Western Shore of Chesapeake Bay.

Little Blue Heron, *Egretta caerulea:* Common in summer along coast from April to October, uncommon to rare inland in spring and late summer. Marshes and impoundments. North Assateague, West Ocean City Pond, Chincoteague, Back Bay, Bombay Hook; Dyke Marsh, Hughes Hollow, Lilypons in late summer.

Tricolored Heron, *Egretta tricolor:* Common from mid-April to mid-October along coast, mainly in Virginia; uncommon to rare in winter. Uncommon in southern Maryland, rare inland. All coastal refuges, Point Lookout; Deal Island in winter.

Cattle Egret, *Bubulcus ibis:* Common, but decreasing, on Delmarva from April to early October, especially with cows, horses, or deer, and near breeding colonies at both ends of Assateague. Uncommon to rare elsewhere. Chincoteague and roads near coast.

Green-backed Heron, *Butorides striatus:* Common and widespread from mid-April to mid-October, rare on Virgina coast in winter, in marshes and swampy woods. C&O Canal, Lake Frank, Dyke Marsh.

Black-crowned Night-Heron, *Nycticorax nycticorax:* Common from mid-April to early November near coast, locally common to rare all year all over coastal plain, rare elsewhere. Coastal refuges, Baltimore Harbor, National Zoo, Conowingo Dam.

Yellow-crowned Night-Heron, *Nyctanassa violacea:* Uncommon and local on Virginia coast and along the Potomac, rare and local elsewhere, from April to September; rare in winter near southeast coast.

Huntley Meadows, Pea Patch Island, Lake Roland (north of Baltimore).

White Ibis, *Eudocimus albus:* Rare and irregular postbreeding visitor, mostly in years of southern drought, on coastal plain and in piedmont; some spring records. Has recently bred in southeast Virginia. Most birds are immatures, in small groups or as singles. More likely in coastal refuges, but has turned up on C&O Canal, at Lilypons, in suburban creeks.

Glossy Ibis, *Plegadis falcinellus:* Common to abundant along Virginia coast, rare farther up Delmarva, from late March to mid-September; uncommon to rare visitor inland in Virginia, mainly mid-April to mid-May. Marshes and muddy fields. Coastal refuges.

Wood Stork,* *Mycteria americana:* Rare and irregular, mostly in summer, in southeast Virginia; casual vagrant in rest of region. Near water.

Fulvous Whistling-Duck,* *Dendrocygna bicolor:* Rare and irregular visitor at all times of year, generally near coast, most often in ponds and freshwater marshes.

Tundra Swan, *Cygnus columbianus:* Common to abundant on Eastern and Western shores from November to April; uncommon to rare migrant and winter visitor inland; rare in summer. Sheltered bays, rivers, ponds; cornfields. Eastern Neck, bays from St. Michaels to Cambridge (Maryland), West Ocean City Pond, Back Bay.

Mute Swan, *Cygnus olor:* Locally common all year, mostly on Delmarva, in sheltered bays, ponds. Single birds and pairs often escape from captivity or are released in urban lakes and ponds. Eastern Neck and Chincoteague refuges, Hooper Island have well-established feral populations; also the pond between the spans at the east end of the Chesapeake Bay Bridge.

Greater White-fronted Goose, *Anser albifrons:* Rare migrant (especially in spring) and winter visitor, found in flocks of wintering Canada Geese. Bombay Hook, Remington Farms, Blackwater, National Geographic Society pond.

Snow Goose, *Chen caerulescens:* Abundant from November to March in coastal refuges. Marshes, sand and mud flats. Bombay Hook, Blackwater, Assateague, Back Bay.

Brant, *Branta bernicla:* Locally common along coast from late October to late April, around jetties, oyster beds, marsh banks, tidal flats. Cape Henlopen, Ocean City, Chincoteague, Fort Monroe (off sea wall).

Canada Goose, *Branta canadensis:* Abundant winter visitor from late September to mid-April on coastal plain, locally common all year, especially from piedmont east. C&O Canal, Merkle Wildlife Management Area, National Geographic Society pond, all Delmarva refuges.

Wood Duck, *Aix sponsa:* Common throughout region from March to mid-November; uncommon on coastal plain in winter, rare in piedmont. Ponds and swampy woods. C&O Canal, Hughes Hollow, National Zoo, Mason Neck.

Green-winged Teal, *Anas crecca:* Common from mid-September to April on coastal plain, uncommon elsewhere. Rare on Delmarva in summer. Dyke Marsh, Hunting Creek, all refuges.

American Black Duck, *Anas rubripes:* Fairly common but decreasing, present all year, especially near coast; uncommon to rare west of coastal plain. Dyke Marsh, Jug Bay, all refuges.

Mallard, *Anas platyrhynchos:* Common to abundant throughout region. Population continually being augmented by released birds. Lake Frank, C&O Canal, duck ponds everywhere.

Northern Pintail, *Anas acuta:* Common from September to April, mainly near coast; often abundant in November and March. Bombay Hook, Chincoteague, Back Bay, Blackwater.

Blue-winged Teal, *Anas discors:* Common from mid-March to April and from August to late October on coastal plain, uncommon elsewhere. Uncommon and local or rare in summer, rare and irregular in winter. Pennyfield Lock, Lilypons, Deal Island, refuges.

Northern Shoveler, *Anas clypeata:* Locally common near coast from mid-September to April, rare inland. Marshes, ponds. Bombay Hook, Blackwater, West Ocean City Pond, Chincoteague.

Gadwall, *Anas strepera:* Common from September to April on coastal plain, uncommon to rare in piedmont. Locally common in summer. Bombay Hook, Deal Island, Chincoteague, Back Bay.

Eurasian Wigeon, *Anas penelope:* Rare and irregular from October to April among flocks of American Wigeon, mostly on Delmarva. Coastal refuges, Remington Farms, Deal Island.

American Wigeon, *Anas americana:* Common but decreasing on coastal plain from mid-September to mid-May, uncommon elsewhere. All refuges, especially Deal Island; ponds, freshwater marshes.

Canvasback, *Aythya valisineria:* Locally common and decreasing, from November to April, mostly on coastal plain; uncommon in piedmont and as migrant in ridge-and-valley province. Sandy Point State Park, West Ocean City Pond, Chesapeake Beach, Georgetown Reservoir, Potomac from Dyke Marsh to Mount Vernon.

Redhead, *Aythya americana:* Uncommon and decreasing on coastal plain from mid-October to April and in piedmont in migration, rare elsewhere. Sometimes in flocks of Canvasbacks in ponds, piedmont lakes, sheltered coves. Baltimore Harbor, Potomac from Dyke Marsh to Mount Vernon, West Ocean City Pond, Chesapeake Beach, Black Hill Park.

Ring-necked Duck, *Aythya collaris:* Common on coastal plain and in piedmont from mid-October to April; common elsewhere in migration, uncommon in winter. Rivers, lakes, and freshwater marshes. Great Falls, Lake Anna, Back Bay, National Geographic Society pond, Georgetown Reservoir.

Greater Scaup, *Aythya marila:* Fairly common on coastal plain from mid-October to April, common in migration; rare elsewhere. Bays, coastal waters (often near jetties), rivers, lakes. Chesapeake Beach, Hooper Island, Ocean City Inlet, Chesapeake Bay Bridge-Tunnel.

Lesser Scaup, *Aythya affinis:* Common on coastal plain from October to April; common elsewhere in migration, uncommon in winter. Ponds, rivers, lakes, and sheltered bays. Baltimore Harbor, Potomac near Wilson Bridge, Lake Anna.

Common Eider, *Somateria mollissima;* **King Eider,** *Somateria spectabilis;* **Harlequin Duck,** *Histrionicus histrionicus:* All three uncommon and local (King Eider somewhat more common) from November to April at coastal jetties. Cape Henlopen, Indian River Inlet, Ocean City Inlet, Chesapeake Bay Bridge-Tunnel, Rudee Inlet.

Oldsquaw, *Clangula hyemalis:* Common from November to April along coast and in lower Chesapeake Bay, uncommon to rare elsewhere. Indian River Inlet, Ocean City Inlet, Chesapeake Bay Bridge-Tunnel, Hooper Island, Point Lookout.

Black Scoter, *Melanitta nigra;* **Surf Scoter,** *Melanitta perspicillata;* **White-winged Scoter,** *Melanitta fusca:* Abundant in migration, common from mid-October to mid-April on coast and in Chesapeake Bay south of the Bay Bridge; rare elsewhere. Assateague, coastal jetties, Chesapeake Bay Bridge-Tunnel, Eastern Neck (in migration), Point Lookout.

Common Goldeneye, *Bucephala clangula:* Common from November to April on coastal plain, uncommon elsewhere. Bays, rivers. Sandy Point, Point Lookout, Eastern Neck, Rehoboth and Indian River bays, Ocean City, Chesapeake Bay Bridge-Tunnel.

Bufflehead, *Bucephala albeola:* Common to abundant from November to April; uncommon west of piedmont. Lakes, rivers, bays, ponds. Potomac River (try Fort Hunt to Mount Vernon), Chesapeake Beach, Bombay Hook, Rehoboth and Indian River bays, Point Lookout.

Hooded Merganser, *Lophodytes cucullatus:* Fairly common to uncommon throughout region (less common on Allegheny plateau),

November to April; rare in summer. Lakes, rivers, sheltered bays. West Ocean City Pond, Chincoteague, Stumpy Lake, Elliott Island, Violette's Lock, mouth of Oxon Run, piedmont lakes.

Common Merganser, *Mergus merganser:* Fairly common but local from November to April on coastal plain and in piedmont. Rivers and lakes. Bombay Hook (Shearness Pool), Conowingo Dam, Potomac (Wilson Bridge, Little Falls, Blockhouse Point, Violette's Lock), Lake Anna.

Red-breasted Merganser, *Mergus serrator:* Common to abundant from November to April in salt water, uncommon to rare elsewhere, rare but regular in summer near coast. Sinepuxent Bay, Assateague, Chesapeake Bay Bridge-Tunnel, Fort Monroe.

Ruddy Duck, *Oxyura jamaicensis:* Locally abundant on coastal plain, uncommon to rare elsewhere from October to April; rare in summer. Rivers, lakes, and impoundments. Potomac (at Four Mile Run and Hunting Creek Bay), Bombay Hook, Chincoteague, sewage ponds.

Black Vulture, *Coragyps atratus:* Fairly common to common in piedmont and coastal plain west and south of Chesapeake Bay; less common in mountains and on Delmarva.

Turkey Vulture, *Cathartes aura:* Common everywhere all year, except rare in mountains in winter and on coastal islands. Open country and upland deciduous woods.

Osprey, *Pandion haliaetus:* Common from April to September on coastal plain; elsewhere uncommon in migration, rare in winter (very rare after December). Coastal bays, Chesapeake Bay, lower Potomac, Assateague, Point Lookout. See Chapter 35.

Bald Eagle, *Haliaeetus leucocephalus:* Locally common resident, mostly on coastal plain. Mason Neck, Pohick Bay, Conowingo Dam, Blackwater, Bombay Hook. See Chapter 35.

Northern Harrier, *Circus cyaneus:* Common on Delmarva from mid-August to April, uncommon from May to mid-August; uncommon elsewhere from September to April, rare rest of year. Marshes and fields; all Delmarva refuges. See Chapter 35.

Sharp-shinned Hawk, *Accipiter striatus:* Rare in summer in highlands, uncommon and widespread in winter. Mainly in deciduous woods and woodland edge. Common to abundant in migration. See Chapter 35.

Cooper's Hawk, *Accipiter cooperii:* Fairly common all year in piedmont and mountains, fairly common winter visitor on coastal plain. Deciduous woods and woodland edge. See Chapter 35.

Northern Goshawk, * *Accipiter gentilis:* Rare in summer in Maryland highlands. Rare throughout region from October to April, in conifers and woodland edge. See Chapter 35.

Red-shouldered Hawk, *Buteo lineatus:* Fairly common (only in appropriate habitat), but perhaps decreasing, all year on Virginia coastal plain (rare north and east of Washington), uncommon all year in piedmont and in winter farther west (where rare in summer). Bottomland deciduous woods and edges. Huntley Meadows, Jug Bay, Myrtle Grove. See Chapter 35. Note: Immatures often mistaken for Broad-winged Hawks.

Broad-winged Hawk, *Buteo platypterus:* Uncommon in summer (rare on coastal plain), common spring migrant, in deciduous woods. Abundant fall migrant. Any sighting from late October to March overwhelmingly unlikely. See Chapter 35.

Red-tailed Hawk, *Buteo jamaicensis:* Common all year in most of region, from September to April on coastal plain, in open country, on woodland edge, in deciduous woods. See Chapter 35.

Rough-legged Hawk, *Buteo lagopus:* Rare to uncommon, locally and irregularly, mostly in northern half of region, from December to March. Open country, both marshes and farmland. Bombay Hook, Elliott Island, Deal Island. See Chapter 35.

Golden Eagle, *Aquila chrysaetos:* Rare throughout region from mid-October to mid-April. Open country, lakes and rivers, usually with Bald Eagles. Highland County, Conowingo Dam, Bombay Hook, Blackwater. See Chapter 35.

American Kestrel, *Falco sparverius:* Common throughout region, especially on coastal plain, from September to April, uncommon for rest of year in open country, mostly in cultivated farmland west of Chesapeake Bay. See Chapter 35.

Merlin, *Falco columbarius:* Rare from late October to early April very close to coast, even more so inland. Common migrant along coast and west side of Chesapeake Bay in April and from mid-September to late October. See Chapter 35.

Peregrine Falcon, *Falco peregrinus:* Fairly common migrant along coast from mid-September to late October, uncommon and local year-round resident and winter visitor. Cape Henlopen lighthouse, water tower at Ocean City, Chincoteague. See Chapter 35.

Ring-necked Pheasant, *Phasianus colchicus:* Uncommon, local, and decreasing introduced species, most widespread in northern part of region, in open country with good cover. Bombay Hook, Lucketts, northern Frederick and Carroll counties (Maryland). Note: The Green

Pheasant, a Japanese subspecies or sister species, was established but appears to be decreasing on southern Delmarva.

Ruffed Grouse, *Bonasa umbellus:* Uncommon all year in mountains and western piedmont. Deciduous and coniferous woodland and edge. Garrett and Allegany counties, Highland County, Shenandoah National Park.

Wild Turkey, *Meleagris gallopavo:* Uncommon all year from western coastal plain to Appalachian plateau. Release programs are successfully increasing the population. Deciduous woods and edges. Riverbend Regional Park, Hughes Hollow and Sycamore Landing Road, Myrtle Grove; Washington and Allegany counties, Prime Hook.

Northern Bobwhite, *Colinus virginianus:* Fairly common but rapidly decreasing resident of piedmont and coastal plain, rare in mountains, especially in Maryland. Open country with good cover; fallow fields and rough meadows, hedgerows and woodland edge. Rock Creek Regional Park, Riverbend Regional Park, Patuxent River Park; Chincoteague.

Yellow Rail,* *Coturnicops noveboracensis:* Rare migrant on coastal plain from late March to May and from late September to November. Fresh and salt marshes. No known sites where it occurs regularly.

Black Rail, *Laterallus jamaicensis:* Uncommon to rare on coastal plain from late April to mid-October. Fresh and salt marshes. Elliott and Deal islands, Broadkill Marsh.

Clapper Rail, *Rallus longirostris:* Common from mid-April to mid-October in salt marshes of coastal plain, uncommon for rest of year. Bombay Hook, Assateague, Deal Island, Chincoteague.

King Rail, *Rallus elegans:* Locally uncommon to rare from April to September, in piedmont and on coastal plain. Fresh and brackish marshes. Huntley Meadows, Remington Farms, Deal Island impoundments.

Virginia Rail, *Rallus limicola:* Common to abundant on coastal plain in summer, uncommon in winter; rare elsewhere. Fresh and brackish marshes. Huntley Meadows, Gunpowder Falls State Park, Sandy Point, Elliott Island, Deal Island, Saxis, Back Bay.

Sora, *Porzana carolina:* Fairly common on coastal plain from mid-April to mid-May and from mid-August to mid-October, especially in September; rare for rest of year. Less common elsewhere. Fresh and brackish marshes. Huntley Meadows, Miller's Island Road, Sandy Point, Jug Bay in migration; Deal Island all year.

Purple Gallinule,* *Porphyrula martinica:* Rare and irregular summer visitor throughout region from April to August; has nested in all three states. Ponds and fresh marshes with abundant water lilies, spatterdock.

Common Moorhen, *Gallinula chloropus:* Uncommon to rare on coastal plain and in piedmont from mid-April to October, rare elsewhere and for rest of year. Fresh marshes; ponds with reeds. Huntley Meadows, Hughes Hollow, Deal Island, Back Bay.

American Coot, *Fulica americana:* Fairly common migrant, uncommon winter visitor from October to early May, more common in southeast; rare in summer. Lakes, rivers, ponds, bays. Georgetown Reservoir, Lake Anna, Deal Island, Back Bay.

Sandhill Crane,* *Grus canadensis:* Rare visitor in open country throughout region at any time of year, most frequently in autumn.

Black-bellied Plover, *Pluvialis squatarola:* Common from August to May near coast, uncommon to rare in June and July; uncommon to rare elsewhere in May, August, and September. Tidal flats, grass flats, ocean beaches. Coastal refuges, Cape Henlopen, Ocean City.

Lesser Golden-Plover, *Pluvialis dominica:* Rare spring migrant, mid-March to mid-May; rare to uncommon fall migrant, late August to early November, mostly on coastal plain. Fallow fields, mud flats, short-grass pasture and turf. Chincoteague, fields and tidal marshes at Bombay Hook.

Wilson's Plover, *Charadrius wilsonia:* Uncommon summer resident on uninhabited Virginia coastal islands; rare, local, and irregular elsewhere along coast from late April to early September. Tidal and sand flats (not ocean beach), often near marsh and dune vegetation. Virginia Coast Reserve.

Semipalmated Plover, *Charadrius semipalmatus:* Abundant migrant near coast, uncommon elsewhere in April and May and from mid-July to October; uncommon to rare on coast for rest of year. Tidal flats, freshwater mud flats, beaches, oyster beds. All coastal refuges.

Piping Plover, *Charadrius melodus:* Uncommon, classified as threatened, from mid-March to mid-October on coast and lower Chesapeake Bay, rare in winter; rare migrant on Western Shore and Potomac beaches. Ocean beaches, sand flats, tidal mud flats. Known breeding areas closed to visitors from April to August. Assateague, Virginia coastal islands.

Killdeer, *Charadrius vociferus:* Common throughout region all year, less common in winter except on coastal plain; more common in migration. Short-grass pasture and turf, plowed fields, mud flats at pond edges. Hunting Creek, Lucketts, Lilypons and vicinity.

American Oystercatcher, *Haematopus palliatus:* Common all year on Delmarva coast from Ocean City south; uncommon and local on

Delaware coast and in lower Chesapeake Bay from March to November. Oyster beds, tidal flats, beaches, salt marshes. Chincoteague, Ocean City (Sinepuxent Bay), Indian River Inlet.

Black-necked Stilt, *Himantopus mexicanus:* Locally common summer resident at Little Creek, Bombay Hook, Deal Island; rare visitor elsewhere near coast from late April to September. Fresh, brackish, and salt marshes. Delaware refuges, Chincoteague, Deal Island.

American Avocet, *Recurvirostra americana:* Locally common near coast, rare elsewhere. Impoundments and salt marshes. Bombay Hook and Little Creek in April and May and from July to October; Chincoteague from August to October; Craney Island from August to February; uncommon to rare for rest of year.

Greater Yellowlegs, *Tringa melanoleuca:* Common on coast in April and May and from mid-July to October, uncommon to rare for rest of year; uncommon inland in migration. Marshes, tidal flats, shallow ponds, impoundments. Delmarva refuges, Lilypons, Hunting Creek.

Lesser Yellowlegs, *Tringa flavipes:* Rare on coast in April and May, but fairly common west of Chesapeake Bay and in Delaware Bay refuges. Common from July to October in marshes, shallow ponds, impoundments. Delmarva refuges, Lilypons, Hunting Creek.

Solitary Sandpiper, *Tringa solitaria:* Fairly common but usually solitary migrant throughout region, uncommon on coast, from late April to late May, and from late July to mid-October. Pond edges, fresh marshes. Meadowside, Pennyfield Lock, Lilypons.

Willet, *Catoptrophorus semipalmatus:* Common summer resident of Delmarva marshes from April to September, rare for rest of year and elsewhere in migration. Salt marshes, ocean beaches, tidal flats. Little Creek, Assateague, Chincoteague, Elliott Island.

Spotted Sandpiper, *Actitis macularia:* Common throughout region in migration, April through May and July through September, uncommon in summer. Lilypons, C&O Canal (river bank), Dyke Marsh.

Upland Sandpiper, *Bartramia longicauda:* Locally uncommon to rare west of coastal plain from April to September; on coastal plain rare in spring, locally uncommon from mid-July to early September. Pastures and hayfields. Lucketts, Garrett County (May through July); Oland Road at Md. 85, Del. 9 at U.S. 113; Greater Wilmington Airport at U.S. 13 and Del. 141 (late July through early September).

Whimbrel, *Numenius phaeopus:* Common to abundant migrant near coast, rare elsewhere on coastal plain, from mid-April to May, and from July to late September; rare on Virginia coast in winter. Ocean beaches and salt marshes. North Assateague, Chincoteague, Virginia Coast Reserve.

Long-billed Curlew,* *Numenius americanus:* Rare visitor to Virginia coastal islands, with most records in fall and winter.

Hudsonian Godwit, *Limosa haemastica:* Rare to locally uncommon fall migrant on coastal plain, regular only at Chincoteague and Bombay Hook (Raymond Pool), mid-July to late October. Shallow fresh, brackish, and salt water, from impoundments to rainpools.

Marbled Godwit, *Limosa fedoa:* Rare in spring (May) and rare to locally uncommon in fall and winter, especially August and September. Shares habitat with Hudsonian Godwit, but more often found on sand flats. Bombay Hook, Little Creek, Chincoteague (regularly), Back Bay. Some winter in Virginia coastal marshes.

Ruddy Turnstone, *Arenaria interpres:* Common to abundant on coast in May, August, September; common at Ocean City and in southeast Virginia in winter; rare farther north; rare inland in migration. Ocean beaches, jetties, tidal flats, muddy fields. Indian River Inlet, Ocean City, Chincoteague, Locustville, Chesapeake Bay Bridge-Tunnel.

Red Knot, *Calidris canutus:* Locally common to abundant migrant on coast from May to early June and from late July to late September; rare for rest of year. Rare in migration elsewhere on coastal plain. Beaches, peat banks, tidal flats; sea walls and jetties, mainly in winter. Chincoteague, Virginia Coast Reserve, Ocean City and Indian River inlets, Port Mahon.

Sanderling, *Calidris alba:* Common on coast in April and May and from late July to September, uncommon in winter, rare in early summer; rare elsewhere. Coastal beaches and tidal and sand flats. Cape Henlopen, Assateague, Virginia Coast Reserve, Back Bay.

Semipalmated Sandpiper, *Calidris pusilla:* Abundant on coast from late April to early June and from mid-July to October, uncommon in early summer and inland. Mud flats, beaches; less likely to use tidal flats and beaches than fresh and brackish mud flats in fall migration. Coastal refuges, Sandy Point.

Western Sandpiper, *Calidris mauri:* Rare in spring, irregularly common in August, common to abundant in September and October, uncommon to rare in winter. Uncommon to rare elsewhere in fall migration. Tidal flats, beaches, fresh and brackish mud flats. Coastal refuges.

Least Sandpiper, *Calidris minutilla:* Common near coast, uncommon elsewhere, from late April to May and from mid-July to late October; uncommon to rare for rest of year on coast. Mud flats, grassy flats, muddy fields. Delmarva refuges, Lilypons, Hunting Creek.

White-rumped Sandpiper, *Calidris fuscicollis:* Uncommon on Eastern Shore, rare elsewhere, from May to early June and from August to early November. Muddy fields and flats. Assateague, Delmarva refuges, Elliott Island.

Baird's Sandpiper, *Calidris bairdii:* Rare from mid-August to late October, mainly on coastal plain. Grass and mud flats, beaches. Delmarva refuges.

Pectoral Sandpiper, *Calidris melanotos:* Rare near coast south of Delaware Bay, uncommon inland and in Delaware refuges, in March and April; common from July to November. Grassy flats, marshes, muddy fields, wet pastures. Delmarva refuges, Lilypons, Hunting Creek.

Purple Sandpiper, *Calidris maritima:* Locally common from November to late May on coastal jetties; rare visitor to Western Shore jetties. Indian River and Ocean City inlets, Chesapeake Bay Bridge-Tunnel; perhaps Sandy Point, Chesapeake Beach, and Point Lookout.

Dunlin, *Calidris alpina:* Abundant on Delmarva from late September to late May, rare from July to mid-September; uncommon to rare inland. Tidal flats, marshes, muddy fields, beaches. Delmarva refuges, especially near coast.

Curlew Sandpiper, *Calidris ferruginea:* Rare (accidental* in Maryland), mainly on coast, in May and from mid-July to October. Mud flats. Bombay Hook, Little Creek, Chincoteague.

Stilt Sandpiper, *Calidris himantopus:* Rare to locally uncommon in spring, uncommon to locally common in fall on Delmarva; rare inland. In April and May and from late July to late October. Coastal refuges, Craney Island.

Buff-breasted Sandpiper, *Tryngites subruficollis:* Rare to irregularly and locally uncommon, mostly near coast, from mid-August to mid-October. Grassy flats, pastures, and turf. Chincoteague, fields near Bombay Hook.

Ruff, *Philomachus pugnax:* Rare, mostly near coast, from late March to mid-May and from July to mid-October. Mud flats, shallow impoundments, rain pools, marshes. Bombay Hook, Little Creek, Chincoteague. Often with yellowlegs.

Short-billed Dowitcher, *Limnodromus griseus:* Abundant on Delmarva from April to early June and from mid-July to October, uncommon to rare in early summer, probably rare in winter (status uncertain because of confusion with Long-billed Dowitcher). Much less common inland on coastal plain, rare elsewhere. Marshes, tidal flats, shallow impoundments. All coastal refuges, Ocean City, north Assateague. Often misidentified as Long-billed Dowitcher.

Long-billed Dowitcher, *Limnodromus scolopaceus:* Rare on coastal plain in April and May; uncommon to fairly common from late July to November; probably rare in winter (status uncertain because of confusion with Short-billed Dowitcher). Mostly fresh and brackish marshes, ponds, and impoundments, sometimes tidal flats. Delmarva refuges, perhaps Lilypons.

Common Snipe, *Gallinago gallinago:* Fairly common in March and April and October and November, common to uncommon in winter on coastal plain, less common elsewhere, especially in highlands. Marshes and wet fields. Dyke Marsh, Accokeek Creek, Blackwater, Chincoteague, Back Bay.

American Woodcock, *Scolopax minor:* Uncommon all year, more common in winter on Virginia coast, less common in mountains; common in migration from mid-March to mid-April and from mid-October to mid-December. Clearings in or near bottomland woods. Hughes Hollow and vicinity, Lilypons, Elliott Island; Eastern Shore National Wildlife Refuge in winter.

Wilson's Phalarope, *Phalaropus tricolor:* Nowhere common, but decreases in numbers from east to west of region. Rare in May and early June, rare to fairly common from late July to October. Fresh and brackish ponds and impoundments, salt marshes. Bombay Hook, Little Creek, Chincoteague, Back Bay, Craney Island.

Red-necked Phalarope, *Phalaropus lobatus:* Uncommon well offshore, locally uncommon on coast, rare inland, from May to early June and from August to mid-October. Open ocean, fresh and brackish ponds and impoundments. Bombay Hook and Little Creek, pelagic trips.

Red Phalarope, *Phalaropus fulicaria:* Uncommon to rare well offshore, very rare inland throughout region, early April to early May, September to October. Shares habitat with Red-necked Phalarope.

Pomarine Jaeger, *Stercorarius pomarinus:* Uncommon well offshore, from late April to May and from September to December. Very rarely seen from Chesapeake Bay Bridge-Tunnel and from coastal beaches.

Parasitic Jaeger, *Stercorarius parasiticus:* Uncommon well offshore, from late April to mid-June and from late August to December. Sometimes seen from Chesapeake Bay Bridge-Tunnel and from coastal beaches.

Long-tailed Jaeger,* *Stercorarius longicaudus:* Rare well offshore in May, August, and September.

Great Skua, *Catharacta skua:* Rare well offshore from December to May.

Laughing Gull, *Larus atricilla:* Abundant (but decreasing north of Virginia) near coast and in Chesapeake Bay, from April to November, rare in winter mainly near mouth of Chesapeake Bay; common along rivers of coastal plain in late summer and fall. Nests in salt marshes near coast. Mud flats, beaches, marshes, pilings. Ocean City, Chincoteague, Sandy Point; in late summer, Back River Sewage Treatment Plant, Hunting Bay.

Franklin's Gull, *Larus pipixcan:* Rare visitor to the coastal plain, mainly west of Chesapeake Bay; records from May to early December, most in May, August, and September, usually with Laughing Gulls. Accidental in Delaware.* Back River, Sandy Point, Hunting Bay.

Little Gull, *Larus minutus:* Rare visitor to coastal plain, with records from August to early June, most in March and early April. Usually with Bonaparte's Gulls, at Back River Sewage Treatment Plant, Indian River Inlet, Fort Story.

Common Black-headed Gull, *Larus ridibundus:* Rare visitor at any time of year, usually with Bonaparte's or Ring-billed Gulls. Back River Sewage Treatment Plant, Indian River Inlet, Ocean City, Chincoteague.

Bonaparte's Gull, *Larus philadelphia:* Common in March and April and from mid-October to November on coastal plain; less common elsewhere. Irregularly common to uncommon in winter, mostly in southeast Virginia. Inlets, harbors, rivers. Back River, Indian River Inlet, Ocean City, Chesapeake Bay Bridge-Tunnel, Fort Story.

Ring-billed Gull, *Larus delawarensis:* Abundant on coastal plain, locally common to abundant in piedmont, uncommon elsewhere from mid-August to early May. Nonbreeders uncommon on coast in summer. The common urban gull in winter, seen in small city parks and shopping malls, and flying around city and suburban streets. The Mall, East Potomac Park, all other gull sites.

Herring Gull, *Larus argentatus:* Common to abundant year-round on coast and from mid-September to mid-May inland on coastal plain; locally common from mid-November to April in piedmont. See previous gull listings for habitats and sites.

Thayer's Gull,* *Larus thayeri;* **Iceland Gull,** *Larus glaucoides;* **Glaucous Gull,** *Larus hyperboreus:* Rare but regular from November to May, on coastal plain and offshore, usually among Herring Gulls. Beaches; sand and mud flats in lagoons, bays, impoundments, and rivers; landfills. Back River Sewage Treatment Plant, Sandy Point, Ocean City, Conowingo Dam, Hunting Bay.

Lesser Black-backed Gull, *Larus fuscus:* Uncommon on coastal plain and in eastern piedmont from September to April; some summer records. See previous gull listings for habitats and sites.

Great Black-backed Gull, *Larus marinus:* Common year-round on coast (nests with Herring Gulls) and from September to April inland on coastal plain; uncommon in piedmont landfills. Hunting Bay, Tidal Basin and East Potomac Park, Baltimore Harbor.

Black-legged Kittiwake, *Rissa tridactyla:* Fairly common well offshore from late October to late March, rarely seen from land. Possible at Ocean City Inlet and Chesapeake Bay Bridge-Tunnel, especially in strong easterly winds.

Gull-billed Tern, *Sterna nilotica:* Rapidly decreasing summer resident on Delmarva coast from May to early September; irregular in Delaware and Maryland, uncommon in Virginia, mainly on uninhabited coastal islands. Rare on shores of lower Chesapeake Bay. Ocean beaches and marshes of coastal islands. Coastal refuges, especially Chincoteague; Virginia Coast Reserve.

Caspian Tern, *Sterna caspia:* Uncommon to locally common in April and May and, especially, from August to October; rare to locally uncommon in summer on coastal plain; rare migrant elsewhere. A rare nester on Virginia coastal islands. Chincoteague, Baltimore County (Miller's Island Road), Hunting Bay, Hog Island.

Royal Tern, *Sterna maxima:* Common from late April to October on coast and Chesapeake Bay; rare in winter in extreme southeast. Nests on coastal islands; gathers in large flocks in late summer and fall; a few wander up tidal rivers in postbreeding period. Cape Henlopen, Chincoteague.

Sandwich Tern, *Sterna sandvicensis:* Uncommon from late April to late September on Virginia coastal islands. Sometimes in flocks of Royal Terns at Chincoteague and Ocean City, usually from late July to early September; Virginia Coast Reserve, Chesapeake Bay Bridge-Tunnel, Point Lookout.

Roseate Tern,* *Sterna dougalli:* Very rare along Virginia coast from April to September, with most records in May. Ocean beaches, tip of Cape Henlopen, Chesapeake Bay Bridge-Tunnel.

Common Tern, *Sterna hirundo:* Common from April to October along coast and Chesapeake Bay, nesting on sand and marsh islands; uncommon on rest of coastal plain, in April and May and in late August and September, rare elsewhere. Most common within a mile of coast, throughout its length. Cape Henlopen, Indian River Inlet, Ocean City, Chincoteague.

Forster's Tern, *Sterna forsteri:* Common from April to November along coast and Chesapeake Bay, nesting on marsh islands; rare to irregularly common in winter in southeast Virginia. Fairly common in late summer and fall along rivers of coastal plain, rare elsewhere. Delmarva refuges and marshes; Hunting Bay, Miller's Island Road, Point Lookout.

Least Tern, *Sterna antillarum:* Common from late April to late September, especially from late July to early August, on coastal plain. Ocean, bay, rivers; nesting on beaches, sandbars, spoil piles, flat rooftops. Coastal parks and refuges, Grandview Preserve, Craney Island.

Bridled Tern,* *Sterna anaethetus:* Rare well offshore in late summer, with most records from mid-August to early September.

White-winged Tern,* *Chlidonias leucopterus:* Very rare visitor to Delaware refuges from Bombay Hook to Ted Harvey, and to Chincoteague, with most recent records in July and August. Usually with Black Terns.

Black Tern, *Chlidonias niger:* Rare in May; uncommon, and decreasing, from July to September along coast; rare migrant inland. Marshes, rivers, bays, ocean. Potomac in May; Bombay Hook, Little Creek, Ted Harvey, Chincoteague.

Black Skimmer, *Rynchops niger:* Common from mid-April to mid-November along coast and lower Chesapeake Bay; uncommon to rare along rivers and upper bay, mainly in late summer and fall; rare in winter in extreme southeast. Nests on coastal islands. Marshes, tidal flats, beaches. Chincoteague, Ocean City, Port Mahon, Sandy Point.

Dovekie,* *Alle alle:* Rare offshore and along coast from mid-November to February.

Razorbill,* *Alca torda:* Rare offshore and along coast, from November to March.

Atlantic Puffin,* *Fratercula arctica:* Rare and irregular well offshore, accidental on coast, from January to March.

Rock Dove, *Columba livia:* Common to abundant all year throughout region. City parks, farmyards, bridges.

Mourning Dove, *Zenaida macroura:* Common all year throughout region (less so in highlands, where uncommon in winter) in woodland edges, hedgerows, fields, gardens.

Common Ground-Dove,* *Columbina passerina:* Very rare visitor, with records from late April to early November. Beaches and sand flats, coastal woodland. Most sightings in southeast Virginia and Maryland's Western Shore.

Black-billed Cuckoo, *Coccyzus erythropthalmus:* Uncommon in migration (most common in mountains), from May to early June and from mid-August to mid-October, throughout region; uncommon in mountains, rare to uncommon in piedmont and coastal plain, from June to mid-August. Deciduous woods and woodland edge. Garrett County parks, Washington County along C&O Canal, Highland County, Shenandoah National Park.

Yellow-billed Cuckoo, *Coccyzus americanus:* Common from late April to October throughout region in deciduous woods and woodland edges. C&O Canal, Seneca Creek State Park, Mason Neck, Huntley Meadows.

Barn Owl, *Tyto alba:* Uncommon all year, decreasing in abundance from east to west of region. Elliott Island. See Chapter 36.

Eastern Screech-Owl, *Otus asio:* Common resident on ridges and in piedmont, uncommon on Appalachian plateau and coastal plain. See Chapter 36.

Great Horned Owl, *Bubo virginianus:* Common throughout region all year. See Chapter 36.

Snowy Owl, *Nyctea scandiaca:* Rare and irregular throughout region from November to March. See Chapter 36.

Barred Owl, *Strix varia:* Locally common resident; rare away from bottomland woods. See Chapter 36.

Long-eared Owl, *Asio otus:* Rare all year except on coastal plain, where present only from mid-November to mid-May. See Chapter 36.

Short-eared Owl, *Asio flammeus:* Local and uncommon from late October to early April in piedmont and on coastal plain, rare in summer on coastal plain; rare from November to April farther west. See Chapter 36.

Northern Saw-whet Owl, *Aegolius acadicus:* Uncommon but rarely seen throughout region from mid-October to mid-March; rare and local in summer in Garrett County and Highland County. See Chapter 36.

Common Nighthawk, *Chordeiles minor:* Locally common throughout region from May to September; abundant from mid-August to early September. Nests on flat city rooftops, migrates in flocks. District of Columbia and suburban centers, Cape Henlopen campground; Potomac River in May; Upton Hill Park, Bull Run Regional Park (Fairfax County, Virginia) in late August.

Chuck-will's-widow, *Caprimulgus carolinensis:* Common on Delmarva, uncommon and local on Maryland's Western Shore, rare visitor to piedmont and mountains, from late April to mid-September.

Pine woods. Piscataway Park, Assawoman, road to Ocean City Airport, Elliott Island Road, Chincoteague.

Whip-poor-will, *Caprimulgus vociferus:* Uncommon and decreasing, throughout region, from April to September; most common in southern Maryland. Rare away from extensive woods. Mason Neck, Blackwater, Pokomoke Swamp.

Chimney Swift, *Chaetura pelagica:* Common throughout region from mid-April to mid-October. In August and September gathers in huge flocks to roost in chimneys and hollow trees. Urban and suburban skies. Upton Hill Park, Metro bus terminal at Wisconsin Avenue and Jenifer Street NW, Washington.

Ruby-throated Hummingbird, *Archilochus colubris:* Fairly common, but decreasing, throughout region from late April to September. Bottomland deciduous woods, gardens, flowering hedgerows and edges. Mason Neck, Mattaponi Creek, C&O Canal, Blackwater.

Belted Kingfisher, *Ceryle alcyon:* Uncommon to fairly common all year throughout region, most common in piedmont. Ponds, creeks, rivers, lakes. C&O Canal, Lilypons, Algonkian Regional Park (Sugarland Run).

Red-headed Woodpecker, *Melanerpes erythrocephalus:* Uncommon and very local in piedmont and on coastal plain in summer, rare and local throughout region in winter in deciduous woods. Huntley Meadows, Mason Neck, Myrtle Grove, Sky Meadows Park (Fauquier County, Virginia), Eastern Neck, Herrington Manor.

Red-bellied Woodpecker, *Melanerpes carolinus:* Common all year throughout region except in highlands. Deciduous woods and woodland edge. C&O Canal, Hughes Hollow, Seneca Creek, Glencarlyn, Huntley Meadows, Oxon Hill, other local parks.

Yellow-bellied Sapsucker, *Sphyrapicus varius:* Fairly common throughout region in April and October, uncommon in winter. Deciduous woods, woodland edge, pine woods, boreal bogs. C&O Canal, Sycamore Landing, Great Falls Park (Virginia), other local parks.

Downy Woodpecker, *Picoides pubescens:* Common throughout region all year in deciduous woods, woodland edge, hedgerows. All local parks.

Hairy Woodpecker, *Picoides villosus:* Uncommon throughout region all year in deciduous woods. C&O Canal, Great Falls Park (Virginia), Jug Bay, other riverside parks.

Red-cockaded Woodpecker, *Picoides borealis:* Very rare, decreasing, and endangered in southernmost Virginia. No publicly known sites.

Northern Flicker, *Colaptes auratus:* Common all year throughout region, especially near coast in fall and winter. Deciduous bottomland woods, woodland edge, open country. C&O Canal, nearby Virginia parks.

Pileated Woodpecker, *Dryocopus pileatus:* Uncommon to locally common throughout region all year. Deciduous woods and pine stands. Glover-Archbold Park, C&O Canal, Great Falls Park (Virginia), Jug Bay, Meadowside.

Olive-sided Flycatcher, *Contopus borealis:* Rare transient throughout region, and in decreasing numbers from west to east, in May and early June and from mid-August to early October. Woodland edges near ponds and swamps, tall snags.

Eastern Wood Pewee, *Contopus virens:* Common throughout region from late April to late October. Deciduous woods and woodland edge. Rock Creek, C&O Canal, Huntley Meadows, Jug Bay.

Yellow-bellied Flycatcher, *Empidonax flaviventris:* Uncommon throughout region from early May to early June and from late August to early October, somewhat more likely in upland deciduous woods, conifers, and swamps.

Acadian Flycatcher, *Empidonax virescens:* Common from late April to mid-September throughout region, generally below 3,500 feet. Deciduous woods often near streams. C&O Canal, Riverbend, Jug Bay, Mason Neck.

Alder Flycatcher, * *Empidonax alnorum:* Uncommon and local nester in western highlands. Arrival and departure dates and abundance of migrants not separated from those of Willow Flycatcher. Nests in alders in boreal bogs usually broken up by young conifers. Cranesville Swamp, Herrington Manor, Finzel Swamp.

Willow Flycatcher, *Empidonax traillii:* Uncommon and local in piedmont west of Chesapeake Bay and in Garrett County from early May to early October. Nests in low wet spots in fields next to marshes, in open marshes and bogs with hummocks or scattered willows and alders. Hughes Hollow, Dyke Marsh, Oxon Hill Farm, Blairs Valley, Cranesville Swamp, Big Meadows.

Least Flycatcher, *Empidonax minimus:* Uncommon throughout region in migration from late April to late May and from late July to late September; common in summer above 2,500 feet (widespread in Garrett County). Typically nests in open second-growth deciduous woods. Swallow Falls, Locust Spring, Shenandoah National Park; Point Lookout, Assateague in fall migration.

Eastern Phoebe, *Sayornis phoebe:* Common throughout region from March to October, uncommon in winter in Virginia. Cliff faces by streams, small bridges, eaves and rafters. Glencarlyn, C&O Canal west of Pennyfield Lock, Meadowside.

Great Crested Flycatcher, *Myiarchus crinitus:* Common throughout region from late April to September. Deciduous woods, pines, woodland edge. Seneca Creek State Park, C&O Canal, Riverbend, Scotts Run, Jug Bay.

Western Kingbird, *Tyrannus verticalis:* Rare in fall, mostly in September and October, very close to coast. A few inland records; sightings as late as January, one or two in May. Bushes, hedgerows, telephone wires in open country and along dune lines.

Eastern Kingbird, *Tyrannus tyrannus:* Common throughout region from late April to September. Woodland edge, hedgerows, open country. Meadowside, Seneca Creek State Park, Hughes Hollow, Huntley Meadows.

Horned Lark, *Eremophila alpestris:* Common all year on coastal plain, uncommon elsewhere. Bare fields; grass and sand flats among coastal dunes. New Design Road, Lucketts, Chincoteague, Cape Henlopen, fields around Bombay Hook.

Purple Martin, *Progne subis:* Locally common throughout region from late March to September. Martin houses are deserted for huge communal roosts in August. Violette's Lock, Bombay Hook, Chincoteague; corner of Pembroke Avenue and Mercury Boulevard in Hampton in late August.

Tree Swallow, *Tachycineta bicolor:* Common to abundant in migration, from April to early May and from late August to October; locally common to rare in summer (from east to west). Ponds, marshes. Hughes Hollow, Lilypons, Blairs Valley, Chincoteague.

Northern Rough-winged Swallow, *Stelgidopteryx serripennis:* Fairly common throughout region from late March to early September. Sandy banks, culverts, bridges. C&O Canal, Lilypons, Mason Neck, Blackwater.

Bank Swallow, *Riparia riparia:* Common in migration on Delmarva from mid-April to mid-May and abundant from mid-July to early September; uncommon elsewhere; locally common at breeding colonies. Sand banks near water. Lilypons, Mason Neck, Jug Bay, Bombay Hook.

Cliff Swallow, *Hirundo pyrrhonota:* Rare in migration on coastal plain from late April to late May and from early July to late September; elsewhere locally common from late April to early September. Barns, bridges, dams. Brighton and Rocky Gorge Dams on Patuxent River.

Barn Swallow, *Hirundo rustica:* Common throughout region, abundant on coastal plain, from April to September. Barns, bridges, house eaves. Riley's Lock, Meadowside, Jug Bay.

Blue Jay, *Cyanocitta cristata:* Common all year throughout region, becoming abundant during migration in late April and May and in October. Hard to avoid except in entirely treeless areas.

American Crow, *Corvus brachyrhynchos:* Common all year throughout region, including all of metropolitan Washington, abundant from late November to late March.

Fish Crow, *Corvus ossifragus:* Common to uncommon in piedmont and on coastal plain, much less common on coast in winter. Nests around Washington Cathedral and Georgetown University. Not separable from Common Crow by location or habitat. Often raids nesting colonies of water birds, as at Cape Henlopen, Deal Island, and Chincoteague, and on coastal and Chesapeake Bay islands.

Common Raven, *Corvus corax:* Local and uncommon all year around Appalachian ridges. Sugarloaf Mountain, Point of Rocks (where U.S. 15 crosses the Potomac), Shenandoah National Park, Highland and Garrett counties, Town Hill.

Black-capped Chickadee, *Parus atricapillus:* Common all year in Maryland from western Washington County west and in Virginia west of Shenandoah Valley from Bath County north in highlands. Infrequent winter visitor from northern states to region in invasion years. (Up to 50 percent of chickadees in central Washington County [Maryland] and Shenandoah Valley are Black-capped × Carolina chickadee hybrids.) Deciduous woods, conifers, woodland edge. Bath and Highland counties (Virginia), Allegany and Garrett counties (Maryland).

Carolina Chickadee, *Parus carolinensis:* Common all year throughout region from eastern Washington County and Blue Ridge east. May be less common than hybrids in central Washington County and Shenandoah Valley. Deciduous woods and woodland edge, pines. Glover-Archbold Park, Glencarlyn, Oxon Hill Farm, and any other local wooded park.

Tufted Titmouse, *Parus bicolor:* Common all year in most of region; rare on coastal islands, uncommon to rare in western highlands. Deciduous woods, woodland edge. All local wooded parks.

Red-breasted Nuthatch, *Sitta canadensis:* Irregularly rare to common throughout region from September to early May, and in mountains all summer. Conifers. National Arboretum, Sandy Point, Assateague.

White-breasted Nuthatch, *Sitta carolinensis:* Common resident in highlands, decreasingly common from west to east, rare on coastal plain. Large trees in deciduous woods.Very quiet and hard to find in breeding season. Great Falls Park (Virginia), C&O Canal, Little Bennett Park.

Brown-headed Nuthatch, *Sitta pusilla:* Locally common all year on the edges of Delmarva south of a line from the Chesapeake Bay Bridge to Cape Henlopen, and in Southside Virginia in loblolly pines. Point Lookout campground, Assawoman, Blackwater, Chincoteague, Seashore State Park, Hog Island.

Brown Creeper, *Certhia americana:* Common in April and October, uncommon to common in winter throughout region, more common in piedmont; locally uncommon but regular and expanding in summer, in Pocomoke Swamp, along the Potomac, and in Garrett County. Deciduous woods, especially bottomland woods. C&O Canal, Sycamore Landing, Riverbend (winter), Washington County (summer).

Carolina Wren, *Thryothorus ludovicianus:* Common all year, in decreasing numbers from east to west, throughout region. Bottomland woods, brush piles and tangles near wooded swamps and streams. Dyke Marsh, C&O Canal, Riverbend, Jug Bay.

House Wren, *Troglodytes aedon:* Locally common throughout region from mid-April to October; rare on coastal plain and in piedmont in winter. Thickets, hedgerows, garden shrubbery, bottomland woods. Sycamore Landing, Meadowside.

Winter Wren, *Troglodytes troglodytes:* Common but hard to find throughout region from October to April; local and uncommon in highlands in summer. Moist woods, near swamps and streamsides. Sycamore Landing, C&O Canal, Great Falls Park (Virginia), Blackwater, Bombay Hook (winter); Highland County, Shenandoah National Park, Cranesville Swamp (summer).

Sedge Wren, *Cistothorus platensis:* Rare, decreasing, and local on Delmarva from May to mid-September (may no longer breed there), more widespread in southeast marshes in winter. Rare and local breeder near bogs in Garrett County. Rare migrant in piedmont. Broomsedge marshes with scattered marsh elder bushes; damp grassy meadows. Elliott and Deal islands; sometimes Bombay Hook, Port Mahon, Broadkill Marsh, Saxis.

Marsh Wren, *Cistothorus palustris:* Abundant in marshes of Delmarva, common along lower Patuxent River, rare and local in piedmont from mid-April to mid-October; uncommon to rare in winter, mostly on Delmarva. Salt and fresh marshes, in fairly extensive patches of tall, coarse grass. Lilypons, Dyke Marsh, Accokeek Creek, Jug Bay, Elliott Island, Bombay Hook.

Golden-crowned Kinglet, *Regulus satrapa:* Common in migration throughout region; common to uncommon in winter on coastal plain and in piedmont, uncommon farther west, from early October to mid-April. Locally common in summer in highlands. Conifers and boreal bogs in summer; conifers and deciduous woods and thickets in migration and winter. Garrett County parks, Cranesville Swamp (summer); National Arboretum, Sandy Point, Blackwater, Chincoteague (winter); riverside parks.

Ruby-crowned Kinglet, *Regulus calendula:* Common throughout region in migration from late March to early May and from late September to November; less common in winter than Golden-crowned Kinglet except in southeast Virginia. Thickets, bottomland woods, pines. Meadowside, riverside parks.

Blue-gray Gnatcatcher, *Polioptila caerulea:* Common throughout region from the end of March to late September; rare in winter in southeast. Open deciduous woods. C&O Canal, Meadowside, Mason Neck, Jug Bay.

Eastern Bluebird, *Sialia sialis:* Uncommon to locally common throughout region all year. Pastures, fields, woodland edges and clearings. Hughes Hollow, Meadowside, Algonkian Park, Merkle.

Veery, *Catharus fuscescens:* Common in highlands in migration, from late April to late May and from late August to September, uncommon farther east; common in summer in the mountains, locally common in Washington, D.C., and Maryland piedmont. Deciduous woods near streams, conifers, bogs. Battery Kemble, Glover-Archbold, and Rock Creek parks, Little Bennett Park, Shenandoah National Park, Garrett County, White Clay Creek, Brandywine Creek.

Gray-cheeked Thrush, *Catharus minimus:* Rare throughout region in migration in May, uncommon in September, October. Habitats and sites as for Swainson's Thrush, in thicker and moister vegetation; much less common.

Swainson's Thrush, *Catharus ustulatus:* Common throughout region in migration in May, September, and October. Woodland, especially along coasts, rivers, ridges. Mason Neck, C&O Canal, Jug Bay, Point Lookout, and all wooded local parks.

Hermit Thrush, *Catharus guttatus:* Common throughout region in migration, in April and from October to early November; fairly common in winter on coastal plain and in Virginia piedmont, uncommon to rare farther west and north. Conifers and deciduous woodland. Fairly common in summer at Herrington Manor and Swallow Falls in Garrett County, rare at Locust Spring in Highland County.

Wood Thrush, *Hylocichla mustelina:* Common throughout region from late April to mid-October. Near forest floor in damp deciduous woods. Scotts Run, C&O Canal, Meadowside, Jug Bay, and all wooded local parks.

American Robin, *Turdus migratorius:* Abundant in migration; common all year on coastal plain (except in north) and in summer farther west; uncommon in winter on northern coastal plain and in piedmont, irregular elsewhere. Gardens, woodland edge, open woods, lawns and fields. Suburban Washington and all local parks.

Gray Catbird, *Dumetella carolinensis:* Common throughout region from late April to early October; rare in winter to uncommon on coastal plain. Hedgerows and gardens, bottomland woods and woodland edge. Sycamore Landing, Hughes Hollow, Seneca Creek, Jug Bay, Huntley Meadows.

Northern Mockingbird, *Mimus polyglottos:* Common all year throughout region except in highlands. Residential city and suburban streets, gardens, hedgerows. Metropolitan Washington, Oxon Hill Farm, National Park Service Nursery, Meadowside, National Arboretum.

Brown Thrasher, *Toxostoma rufum:* Formerly common, now decreasing, throughout region from April to mid-October; rare in winter, rare to uncommon on coastal plain. Hedgerows, tangles, and thickets in outer suburbs, open country. Meadowside, Sycamore Landing, near Potomac west of Washington County.

American Pipit, *Anthus rubescens:* Common on coastal plain, especially in southeast, from October to April; uncommon and irregular in rest of region, chiefly in migration from mid-March to April and from October to early November. Plowed and harvested fields, mud flats. Fields near Point Lookout, tip of Delmarva (winter), New Design Road (migration).

Cedar Waxwing, *Bombycilla cedrorum:* Common in mountains, uncommon in piedmont and on coastal plain, except in mid-winter, when rare away from coast. Woodland edge, clearings, hedgerows, gardens and orchards, wherever there are small fruits and berries. Meadowside, National Arboretum (the cherry grove), Indian Springs.

Northern Shrike,* *Lanius excubitor:* Very rare throughout region, with records from late October to early March. Woodland edge, hedgerows by fields, marshes.

Loggerhead Shrike, *Lanius ludovicianus:* Uncommon resident in Shenandoah Valley, southern Frederick Valley, and northwestern Virginia piedmont. Lilypons, Lucketts.

European Starling, *Sturnus vulgaris:* Abundant throughout region all year.

White-eyed Vireo, *Vireo griseus:* Common on coastal plain and in piedmont from mid-April to mid-October; uncommon to locally common farther west, mostly in lowlands. Thickets and tangles on woodland edge, hedgerows. Seneca Creek, Meadowside, Hughes Hollow, Jug Bay.

Solitary Vireo, *Vireo solitarius:* Uncommon on coastal plain and in piedmont in migration, from early April to early May and from late September to October; common in highlands from early April to October; rare in southern Virginia in winter. In conifers and mixed woods (summer); deciduous and mixed woods (migration). Shenandoah National Park; Highland, Allegany, and Garrett counties; riverside parks in migration.

Yellow-throated Vireo, *Vireo flavifrons:* Common in piedmont, uncommon to locally common elsewhere from mid-April to mid-October. Crowns of tall deciduous trees beside streams or roads, woodland edge. Violette's Lock, Riley's Lock, Algonkian Park, Great Falls Park (Virginia), White Clay Creek.

Warbling Vireo, *Vireo gilvus:* Uncommon to locally common in piedmont and in lowlands farther west from late April to September, rare transient elsewhere. Mature deciduous trees overhanging water. C&O Canal, especially from Violette's Lock to Sycamore Landing and in Washington and Allegany counties.

Philadelphia Vireo, *Vireo philadelphicus:* Very rare throughout region in May, a little more common in September. Woodland edge, along streams, fields, roads, clearings. C&O Canal (especially Violette's Lock), Shenandoah National Park, Assateague.

Red-eyed Vireo, *Vireo olivaceus:* Abundant throughout region from mid-April to October. Deciduous woods. All wooded local parks.

Blue-winged Warbler, *Vermivora pinus:* Uncommon to rare on coastal plain and in piedmont from late April to early May and from late August to mid-September. Local and uncommon in ridge-and-valley province and throughout Maryland piedmont from late April to mid-September. Overgrown clearings and fields with young trees, woodland edge. Little Bennett Park; in migration, C&O Canal, Algonkian Park.

Golden-winged Warbler, *Vermivora chrysoptera:* Rare on coastal plain and in piedmont from end of April to mid-May and from early August to mid-September; uncommon and local in western mountains from mid-May to mid-September. Woodland edge by overgrown fields, clearings, streams. Highland, Washington, Allegany, and Garrett counties; in migration, any local park, especially C&O Canal.

Tennessee Warbler, *Vermivora peregrina:* Rare on coastal plain, rare to uncommon in piedmont, uncommon in mountains in May; somewhat more common in each province from the end of August to mid-October. Deciduous woods and woodland edge. Riverside parks, Shenandoah National Park.

Orange-crowned Warbler, *Vermivora celata:* Very rare throughout region in May, rare but perhaps increasing in late fall and early winter, especially on coast; a few winter on Delmarva. Thickets, hedgerows, woodland edge.

Nashville Warbler, *Vermivora ruficapilla:* Rare on coastal plain, rare to uncommon in piedmont, uncommon in mountains in first half of May and from early September to mid-October. Uncommon and local in summer in Garrett county. Boreal bogs, woodland edge and scrub. Cranesville Swamp (summer), local parks.

Northern Parula, *Parula americana:* Common throughout region from mid-April to mid-October. Bottomland woods, close to water. C&O Canal, Great Falls Park (Virginia), Riverbend, Mason Neck, Jug Bay, North Pocomoke Swamp, Dismal Swamp.

Yellow Warbler, *Dendroica petechia:* Common on Delmarva, uncommon to common in rest of region, from mid-April to mid-September; most common in first half of May. Hedgerows, thickets, willows and alders, near water, marshes, and wet meadows. Hughes Hollow, Lilypons, Blairs Valley, all Delmarva refuges.

Chestnut-sided Warbler, *Dendroica pensylvanica:* Very rare on Delmarva in spring, uncommon in fall; common farther west from end of April to late May and from late August to late September; common in western highlands in summer. Woodland edge. Shenandoah National Park, Highland and Garrett counties, riverside parks.

Magnolia Warbler, *Dendroica magnolia:* Fairly common throughout region from end of April to late May and from late August to mid-October; uncommon in western highlands in summer. Conifers near water in summer; deciduous and mixed woods in migration. Locust Spring, Herrington Manor, Swallow Falls (summer), riverside parks.

Cape May Warbler, *Dendroica tigrina:* Uncommon on coastal plain and in piedmont and common farther west from the end of April to late May; common throughout region from September to mid-October. Conifers and deciduous trees along woodland edge and in open forest. Local parks.

Black-throated Blue Warbler, *Dendroica caerulescens:* Fairly common throughout region from late April to late May and from September to mid-October. Common in western highlands in summer. Mixed woodlands with dense understory; laurel and rhododendron thickets; also woodland edge and hedgerows in migration. Shenandoah National Park, Locust Spring, Herrington Manor, Swallow Falls, local parks.

Yellow-rumped Warbler, *Dendroica coronata:* Common throughout region in migration, abundant in winter on Delmarva, uncommon elsewhere, October to mid-May. Rare in summer in Garrett County.

Thickets of wax myrtle, red cedar, poison ivy in winter; woodland edge and hedgerows in migration. All Delmarva refuges, local parks.

Black-throated Green Warbler, *Dendroica virens:* Uncommon on coastal plain, common elsewhere from mid-April to mid-May and from September to mid-October. Locally common nester in Dismal and Pokomoke swamps from end of March to early July and in mountains (especially in hemlocks) from mid-April to October. Shenandoah National Park, Ramsey's Draft, Highland County, Green Ridge, Garrett County parks, riverside parks.

Blackburnian Warbler, *Dendroica fusca:* Rare on coastal plain in spring and uncommon in fall; uncommon in piedmont and mountains and valleys in first three weeks of May and from mid-August to mid-October; uncommon in summer in western highlands. Mixed woodland with tall trees, especially conifers. Shenandoah National Park, Locust Spring, Garrett County, C&O Canal; Great Falls Park (Virginia) in migration.

Yellow-throated Warbler, *Dendroica dominica:* Common on coastal plain, locally common to uncommon in piedmont, locally uncommon to rare in western lowlands as far as Cumberland. Deciduous bottom-land woods with tall sycamores, cypress, mature pines. C&O Canal, especially west of Riley's Lock, Myrtle Grove, Jug Bay, North Pocomoke Swamp, Dismal Swamp.

Pine Warbler, *Dendroica pinus:* Common on coastal plain and in southern piedmont, locally common to rare farther west and north, from early March to late October. Uncommon to rare in southeast, rare and irregular elsewhere in winter. Pine woods and woodland edge. Huntley Meadows, National Arboretum, Seneca Creek State Park, Sandy Point.

Prairie Warbler, *Dendroica discolor:* Common throughout region except in highlands (where uncommon), from mid-April to late September; a few sometimes linger into December. Overgrown fields with scattered junipers, hedgerows, woodland edge. Jug Bay, Huntley Meadows, Seneca Creek State Park, Meadowside.

Palm Warbler, *Dendroica palmarum:* Uncommon throughout region from early April to early May and west of coastal plain from early September to October. Common on coastal plain, mainly close to coast, from mid-September to October, sometimes much later in southeast Virginia. Very rare elsewhere in winter. Woodland edge and scrub near water, muddy fields, grassy flats, dunes. Riverside parks, coastal refuges.

Bay-breasted Warbler, *Dendroica castanea:* Rare in spring, uncommon in fall on coastal plain; uncommon elsewhere from early to late May; somewhat more common from end of August to mid-October.

Deciduous woods and woodland edge. Parks in Fairfax, Loudoun, and Montgomery counties.

Blackpoll Warbler, *Dendroica striata:* Common throughout region from May to early June, somewhat less common from mid-September to late October. Deciduous woods, woodland edge. Local parks.

Cerulean Warbler, *Dendroica cerulea:* Locally common west of coastal plain from late April to early September. Open deciduous woods with tall trees near streams; woodland edge. C&O Canal; Great Falls Park (Virginia); Shenandoah National Park; Highland, Washington, and Allegany counties; White Clay Creek.

Black-and-White Warbler, *Mniotilta varia:* Fairly common throughout region from early April to mid-October, less common in summer in the south; some winter records. Deciduous and mixed woods. Mason Neck, Great Falls Park (Virginia), Shenandoah National Park, North Pocomoke Swamp, all local parks in migration.

American Redstart, *Setophaga ruticilla:* Common throughout region, especially in the mountains from late April to early October, though local in the east. Open deciduous woods with second-growth trees or well-grown understory, especially in swampy areas. C&O Canal, Jug Bay, Mason Neck.

Prothonotary Warbler, *Protonotaria citrea:* Locally common on coastal plain and in piedmont along the Potomac from mid-April to mid-September, rare elsewhere. Bottomland woods and swamps. C&O Canal, especially from Fletcher's Boat House to Carderock; Myrtle Grove, North Pocomoke Swamp, Dismal Swamp.

Worm-eating Warbler, *Helmitheros vermivorus:* Fairly common locally on coastal plain, uncommon in piedmont and in western river valleys, common on slopes of western ridges, from late April to late September. Hillsides in deciduous woods with scattered rhododendrons, often above swamps and streams. North Pocomoke Swamp, Dismal Swamp, Great Falls Park (Virginia), Mason Neck, Shenandoah National Park, Highland and Washington counties.

Swainson's Warbler, * *Limnothlypis swainsonii:* Locally common in Southside Virginia; rare and local on Delmarva, mid-April to early September; rare visitor elsewhere. Canebrakes and catbrier tangles in cypress swamps and pine woods; swampy understory in bottomland woods, rhododendron thickets. Dismal Swamp.

Ovenbird, *Seiurus aurocapillus:* Common throughout region from mid-April to mid-October. Upland deciduous woods and pines, on forest floor. Scotts Run, Great Falls Park (Virginia), C&O Canal, Jug Bay, Sugarloaf Mountain.

Northern Waterthrush, *Seiurus noveboracensis:* Uncommon throughout region from late April to late May and from early August to early October; nests in Garrett County. Woodland swamps, bogs, streams. Swallow Falls, Cranesville and Finzel swamps, riverside parks, Meadowside, Mason Neck.

Louisiana Waterthrush, *Seiurus motacilla:* Common throughout region from end of March to mid-September. Bottomland forest streams. C&O Canal, Great Falls Park (Virginia), Jug Bay, Myrtle Grove, North Pocomoke Swamp.

Kentucky Warbler, *Oporornis formosus:* Common on coastal plain, locally common in piedmont, uncommon farther west, from end of April to early September. Moist bottomland forest with dense understory. Great Falls Park (Virginia), C&O Canal, Jug Bay, Mason Neck, Myrtle Grove, North Pocomoke Swamp.

Connecticut Warbler,* *Oporornis agilis:* Rare throughout region from September to mid-October. Damp woodlands and woodland edge in brush, tangles; tall weeds.

Mourning Warbler, *Oporornis philadelphia:* Uncommon in ridge-and-valley province from early May to early June and from August to September, rare farther east. Rare and local in summer in western highlands. Dense brush and brier tangles by roadsides, bogs, woodland edge. Western Highland County, Table Rock on Backbone Mountain in Garrett County.

Common Yellowthroat, *Geothlypis trichas:* Abundant on coastal plain, common elsewhere mid-April to late October, rare in winter. Thickets, tangles, and cattail stands in and around marshes, wet meadows. Hughes Hollow, Dyke Marsh, Jug Bay, Meadowside.

Hooded Warbler, *Wilsonia citrina:* Locally common throughout region except in piedmont, from late April to mid-September. Swampy bottomland woods; in mountains, forested slopes, usually near streams. Garrett County parks, Myrtle Grove, Jug Bay, North Pocomoke Swamp, Dismal Swamp.

Wilson's Warbler, *Wilsonia pusilla:* Rare on coastal plain and in southern piedmont, uncommon elsewhere from early to late May and from late August to late September. Hedgerows and thickets near streams and wet meadows. Hughes Hollow, Lilypons, Washington County.

Canada Warbler, *Wilsonia canadensis:* Rare on coastal plain (more common in fall), uncommon to common in piedmont, and common farther west, from early to late May and from mid-August to late September; common in highlands in summer. Woodland stream and swamp edges in summer; clearings and thickets in migration. Shenandoah National Park, Locust Spring, Swallow Falls, Meadowside, Great Falls Park (Virginia), other riverside parks.

Yellow-breasted Chat, *Icteria virens:* Uncommon and decreasing throughout region, from late April to late October. Rare on coastal plain and very rare in piedmont in winter. Hedgerows and tangles in cutover forest, abandoned fields, overgrown pastures. Hughes Hollow, Seneca Creek State Park, Meadowside, Washington County, Jug Bay.

Summer Tanager, *Piranga rubra:* Fairly common on southern coastal plain and in southern piedmont, uncommon to rare and local elsewhere, from late April to late September. Open mixed woodland of pine, oak, and hickory. Great Falls Park (Virginia), Jug Bay, Easton, Blackwater, Elliott Island Road, North Pocomoke Swamp. (Much more likely at Delmarva sites.)

Scarlet Tanager, *Piranga olivacea:* Common throughout region from late April to early October, in deciduous woods. C&O Canal, Great Falls Park (Virginia), Riverbend, Jug Bay, Sugarloaf Mountain.

Western Tanager,* *Piranga ludoviciana:* Very rare, with most records from coastal plain, from October to May.

Northern Cardinal, *Cardinalis cardinalis:* Common to abundant throughout region all year. Urban and suburban residential areas, bottomland deciduous woods, woodland edge, hedgerows. Metropolitan Washington, local parks.

Rose-breasted Grosbeak, *Pheucticus ludovicianus:* Rare to uncommon on coastal plain; uncommon in piedmont, uncommon to common farther west, from late April to late May, and from late August to late October. Locally common in highlands in summer. Woodland edge with large deciduous trees and tall shrubs. Shenandoah National Park, Highland County, Garrett County parks in summer; C&O Canal, Great Falls Park (Virginia) in migration.

Black-headed Grosbeak,* *Pheucticus melanocephalus:* Very rare visitor throughout region, with records from October to May, most of them at feeders.

Blue Grosbeak, *Guiraca caerulea:* Common on coastal plain, especially Delmarva; fairly common in piedmont, uncommon in lowlands farther west, from May to late October. Woodland edge and hedgerows in farmland. Sandy Point, Jug Bay, Elliott Island Road, farmland around Leipsic, North Pocomoke Swamp, W&OD Trail; Point Lookout in autumn.

Indigo Bunting, *Passerina cyanea:* Common to abundant throughout region from late April to late October. Woodland edge, scattered trees in pastures, fields, clearings. Hughes Hollow, Seneca Creek State Park, Meadowside, Jug Bay, Merkle.

Painted Bunting,* *Passerina ciris:* Very rare, with scattered records from August to May, mostly on coastal plain and at feeders.

Dickcissel, *Spiza americana:* Rare and irregular throughout region, mainly in northern piedmont, from May to September; in winter, rare but regular in southeast Virginia, elsewhere occasionally reported at feeders. Alfalfa fields. In last decade five or six pairs a year reported on territory in Carroll (Maryland), Frederick (Maryland), and Loudoun (Virginia) counties.

Rufous-sided Towhee, *Pipilo erythrophthalmus:* Common throughout region; from November to March more common on southeast coastal plain, much less common in northern piedmont and western valleys, absent from highlands. Thickety woodland edge and hedgerows. Seneca Creek State Park, Meadowside, Riverbend, Huntley Meadows, Jug Bay.

Bachman's Sparrow,* *Aimophila aestivalis:* Rare, local, and endangered in southernmost Virginia. No publicly known sites.

American Tree Sparrow, *Spizella arborea:* Uncommon to rare throughout region from early November to late March; more common north of Washington, irregular south of it. Weedy fields near thickets and woods. Fields off River Road, Black Hill Park, Lilypons, Sandy Point, Bombay Hook, Brandywine Creek Park.

Chipping Sparrow, *Spizella passerina:* Common throughout region from late March to early November; rare in winter, sometimes uncommon in southeast Virginia. Suburban parks and gardens, shaded lawns and pastures, grassy clearings. Fort Hunt, Belle Haven Picnic Area, Montrose Park, Jug Bay, Meadowside.

Clay-colored Sparrow, *Spizella pallida:* Rare but regular visitor, with almost all records along coast, from September to November. Usually with Field Sparrows. Grassy areas. Assateague.

Field Sparrow, *Spizella pusilla:* Fairly common throughout region all year, but decreasing; more common on coastal plain in winter. Weedy fields, thickets, hedgerows, woodland edge. Hughes Hollow, Meadowside, Jug Bay, Huntley Meadows.

Vesper Sparrow, *Pooecetes gramineus:* Uncommon throughout region from early March to early May and from late September to early November; rare on coastal plain and in southern piedmont in summer; uncommon to fairly common elsewhere; rare in winter. Upland pastures and fields. Lucketts, New Design Road. Sussex County (Delaware) fields.

Lark Sparrow, * *Chondestes grammacus:* Rare visitor along coast, with most records in August and September, a few in May, and others from October to April. Even rarer in highlands and piedmont. Grassy areas by coastal dunes; pastures. Assateague, fields east of Back Bay.

Savannah Sparrow, *Passerculus sandwichensis:* Common in winter along coast, uncommon to rare inland; common throughout region from mid-March to mid-May and from mid-September to early November. Uncommon in summer in northern piedmont and western highlands. Wet meadows, grassy flats next to fresh or tidal marshes. Ipswich race in sandy coastal dunes. Delmarva refuges in winter; Jug Bay, Sandy Point in migration; Blairs Valley, Highland and Garrett counties in summer.

Grasshopper Sparrow, *Ammodramus savannarum:* Uncommon to locally common throughout region from mid-April to late October. Hay fields and old pastures. Lucketts, Seneca Creek State Park, New Design Road.

Henslow's Sparrow, * *Ammodramus henslowii:* Local, rare, and decreasing throughout region (perhaps extirpated from Eastern Shore), from mid-April to late October. Weedy meadows, predominantly broomsedge, often at marsh edges; reclaimed strip mines. Locations not stable from year to year, but regular in Garrett County and eastern Allegany County (Maryland).

Sharp-tailed Sparrow, *Ammodramus caudacutus:* Locally common on coast and Chesapeake Bay in migration and in Maryland Eastern Shore marshes in summer; common on southern Delmarva in winter. Salt and fresh marshes. Rare in wet weedy fields and fresh marshes in migration. Deal and Elliott islands (all year), Assateague and Chincoteague, Saxis (winter).

Seaside Sparrow, *Ammodramus maritimus:* Common to abundant in Delmarva marshes, regular and uncommon in Chesapeake Bay, from late April to early November; uncommon to rare in winter, more numerous on Virginia Eastern Shore. Salt marshes. Port Mahon, Elliott Island, Assateague and Chincoteague, Deal Island.

Fox Sparrow, *Passerella iliaca:* Fairly common throughout region from late February to March, uncommon in November and in southeast Virginia in winter; uncommon to rare in rest of region in winter. Forest floor in mixed woods and woodland edge near thickets and brush piles. Hughes Hollow, Jug Bay, local wooded parks; Norfolk area parks in winter.

Song Sparrow, *Melospiza melodia:* Common throughout region all year (less so on southern coastal plain), abundant on coastal plain and in piedmont in winter. Hedgerows, thickets, brush piles, weedy fields, swamps, shrubbery. Residential areas, Meadowside, Hughes Hollow, Oxon Hill Farm, National Arboretum.

Lincoln's Sparrow, *Melospiza lincolnii:* Uncommon throughout region in April and May and from late September to early November; rare in southeast Virginia in winter. Brush piles and thickets at edges of weedy fields and marshy areas.

Swamp Sparrow, *Melospiza georgiana:* Uncommon to abundant throughout region in migration, increasing in numbers from northwest to southeast; common to abundant on coastal plain and in southern piedmont in winter, rare to uncommon in northern piedmont and western valleys. Uncommon and local in summer in western highlands and Delmarva. Marshes with rank vegetation, weedy fields, hedgerows. Hughes Hollow, Dyke Marsh, Miller Island Road, Jug Bay, Sandy Point, Bombay Hook (all year); Locust Spring, Cranesville Swamp (summer).

White-throated Sparrow, *Zonotrichia albicollis:* Common to abundant throughout region from October to late May, rare in Maryland highlands in summer.

White-crowned Sparrow, *Zonotrichia leucophrys:* Uncommon east of mountains, common in western valleys from early October to early May. Multiflora and other hedgerows, thickets, weedy fields. Lilypons, Lucketts, Shenandoah Valley, Remington Farms.

Harris's Sparrow,* *Zonotrichia querula:* Rare visitor throughout region except for Virginia coastal plain (no records), with records from October to March. Brush piles, hedgerows, weedy fields, but especially at feeders.

Dark-eyed Junco, *Junco hyemalis:* Common to abundant throughout region from October to April; common in Virginia mountains in summer, much less common in Maryland mountains. Conifers and deciduous woods, along roads and clearings; urban and suburban residential areas, hedgerows, and field edges in winter. Shenandoah National Park, Highland County in summer; metropolitan Washington and local parks in winter, especially nature centers with feeders.

Lapland Longspur, *Calcarius lapponicus:* Rare near coast, rare and irregular inland in northern third of region from late October to April, usually seen from December to February. Grassy mud flats, dunes, harvested fields, turf, plowed road edges after a snowfall; usually with Horned Larks. Kent County (Delaware and Maryland), Lower Frederick County, Cape Henlopen, Assateague.

Snow Bunting, *Plectrophenax nivalis:* Uncommon near coast, rare and increasingly irregular from east to west, from end of October to March. Dunes, sand and mud flats, harvested fields. Often with Horned Larks. Cape Henlopen, Assateague, lower Frederick County.

Bobolink, *Dolichonyx oryzivorus:* Common throughout region from early to late May and, especially on coast, from August to early October; locally common in Highland and Garrett counties in summer. Fields of alfalfa, clover, and hay in spring and summer, reed marshes in fall. Highland County south of Bluegrass, Garrett County south of Oakland; alfalfa fields near Lucketts, New Design Road, western Montgomery County in spring; Bombay Hook and Little Creek in fall.

Red-winged Blackbird, *Agelaius phoeniceus:* Common throughout region all year; abundant on coastal plain, irregular elsewhere in winter. Fresh and brackish marshes, fields, pastures, salt marshes, woodlots. Dyke Marsh, Hughes Hollow, Lilypons, Delmarva.

Eastern Meadowlark, *Sturnella magna:* Locally common throughout region all year, but decreasing; far more common on Delmarva in winter. In meadows, fields, pastures, and, locally, in marshes. More likely along country roads than in refuges and parks. Lucketts, New Design Road.

Yellow-headed Blackbird,* *Xanthocephalus xanthocephalus:* Rare and irregular, mostly on Delmarva, from September to April. Harvested fields, pastures, and marshes; usually a single bird in huge flocks of Red-winged Blackbirds, grackles, and cowbirds.

Rusty Blackbird, *Euphagus carolinus:* Uncommon throughout region from mid-October to late April, somewhat more common before mid-November and after mid-March. Swampy woods, and marsh edges near trees on coast. Rarely in big blackbird flocks. Dyke Marsh, Oxon Hill Farm, Seneca Marsh (above Riley's Lock), Hughes Hollow, Sycamore Landing Road, Myrtle Grove.

Brewer's Blackbird,* *Euphagus cyanocephalus:* Rare throughout region from October to April; most records from coastal plain and western Virginia valleys. In harvested fields, farmyards, pastures; usually not in large flocks with other blackbirds. Bombay Hook and vicinity, Nokesville area of Prince William County; Back Bay vicinity.

Boat-tailed Grackle, *Quiscalus major:* Common all year on Virginia and Maryland coast and, locally, in lower Chesapeake Bay. Common in summer but less common in winter on Delaware coast. Small breeding colony at Point Lookout. Salt marshes. Coastal refuges, Ocean City (winter), Point Lookout.

Common Grackle, *Quiscalus quiscula:* Common throughout region, abundant in winter on coastal plain, irregularly abundant elsewhere. Fields, swampy woods, conifers. All local parks and farmland.

Brown-headed Cowbird, *Molothrus ater:* Common throughout region from April to November. Locally common in winter west of

Chesapeake Bay. Woodland edge, woods, harvested fields. All local parks and farmland.

Orchard Oriole, *Icterus spurius:* Common on coastal plain and in piedmont, common to uncommon in lowlands farther west, from May to August. Scattered shade trees in open country, orchards, farmyards. Seneca Creek State Park, Meadowside, Huntley Meadows, Pennyfield Lock, Riley's Lock, Lilypons, Gunpowder Falls State Park.

Northern (Baltimore) **Oriole,** *Icterus galbula:* Common in northern piedmont and lowlands farther west from late April to September; uncommon on coastal plain, common in southern piedmont from late April to late May, common throughout from mid-August to September. Rare in winter, often at feeders. Woodland edge, shade trees in open country and near houses, often near water. Fletcher's Boat House, C&O Canal, Meadowside, Algonkian Park, Jug Bay, Lilypons.

Purple Finch, *Carpodacus purpureus:* Irregular and very uncommon throughout region from October to early May, more common from late April to early May, and from late September to early October. Uncommon to rare, and decreasing, in summer in western highlands. Mixed coniferous and deciduous woods. Garrett County parks in summer; C&O Canal (especially Violette's Lock to Riley's Lock), Great Falls Park (Virginia), Mason Neck, Jug Bay, National Arboretum in winter and in migration.

House Finch, *Carpodacus mexicanus:* Abundant and rapidly increasing throughout region from mid-October to early May; uncommon to common but expanding throughout region in summer. Urban and suburban areas. Metropolitan Washington.

Red Crossbill, *Loxia curvirostra:* Irregular and rare in most of region; apparently a permanent resident on Shenandoah Mountain. Usually seen and heard in flight over stands of conifers.

White-winged Crossbill, *Loxia leucoptera:* Rare and very irregular throughout region from November to May. Conifers. More often in residential areas than Red Crossbill.

Common Redpoll, *Carduelis flammea:* Rare and increasingly irregular in northern half of region from December to mid-April. Weedy fields, birches, alders, coastal dunes, feeders.

Pine Siskin, *Carduelis pinus:* Irregularly rare to common throughout region, mid-September to mid-May, most often from late September to early October and late April to early May. Conifers, weeds, alders, woodland edge, feeders. Often with goldfinches. National Arboretum, Chinquapin Park, feeders.

American Goldfinch, *Carduelis tristis:* Common throughout region all year; abundant from early April to early May and from mid-October to mid-November. Weedy fields, hedgerows, tangles on woodland edge; feeders. C&O Canal, Hughes Hollow, Meadowside, Jug Bay, Huntley Meadows.

Evening Grosbeak, *Coccothraustes vespertinus:* Usually rare to occasionally abundant throughout most of region from late October to mid-May, most likely from late April to early May; regular and fairly common in Garrett County in winter. Mostly seen at feeders with sunflower seeds, sometimes in box elders. Feeding stations in Rock Creek Park, Meadowside, Riverbend.

House Sparrow, *Passer domesticus:* Common throughout region all year, near people.

RARITIES AND ACCIDENTALS

The following species have been recorded in the region from 1960 to 1990, but with such infrequency that even very active birders should not expect to see them in five years or more of intensive and varied field work. Several have been seen only once. Species regarded by the current state records committees as inadequately documented by present standards have not been included.

Sightings of birds listed below or marked by an asterisk (*) on the preceding pages, or not mentioned at all, deserve good documentation, including photographs when possible, a prompt effort to notify other birders, and a report to the state records committee (see Appendix D) with a copy to the appropriate regional editor of *American Birds.*

Pacific Loon	Black-shouldered Kite	Arctic Tern
Yellow-nosed Albatross	Mississippi Kite	Bridled Tern
Western Grebe	Swainson's Hawk	Sooty Tern
White-tailed Tropicbird	Gyrfalcon	Common Murre
Black-capped Petrel	Paint-billed Crake	Thick-billed Murre
White-faced Storm-Petrel	Greater Golden-Plover	Black Guillemot
Brown Booby	Mountain Plover	White-winged Dove
Magnificent Frigatebird	Rufous-necked Stint	Groove-billed Ani
White-faced Ibis	Little Stint	Burrowing Owl
Roseate Spoonbill	Sharp-tailed Sandpiper	Rufous Hummingbird
Greater Flamingo	South Polar Skua	Lewis's Woodpecker
Barnacle Goose	Mew Gull	Three-toed Woodpecker
White-cheeked Pintail	California Gull	Black-backed Woodpecker
Garganey	Ross's Gull	Western Wood-Pewee
Limpkin	Sabine's Gull	Hammond's Flycatcher
American Swallow-tailed	Ivory Gull	Say's Phoebe
Kite	Elegant Tern	Ash-throated Flycatcher

Gray Kingbird
Scissor-tailed Flycatcher
Fork-tailed Flycatcher
Boreal Chickadee
Rock Wren
Bewick's Wren
Northern Wheatear
Mountain Bluebird
Fieldfare
Varied Thrush

Sage Thrasher
Sprague's Pipit
Bohemian Waxwing
Bachman's Warbler
Black-throated Gray
   Warbler
Kirtland's Warbler
Lazuli Bunting
Green-tailed Towhee
Black-throated Sparrow

Lark Bunting
Baird's Sparrow
Le Conte's Sparrow
Smith's Longspur
Chestnut-collared
   Longspur
Pine Grosbeak
Hoary Redpoll

Washington and Vicinity

The District of Columbia   4

Washington's abundance of wooded parks and open and sheltered waters means for the hurried or car-less birder that there is no need to go beyond the city limits to see an impressive variety of birds. In recent years diligent field observers have shown that some 240 species can be found in a year of single-minded searching if no key habitats are neglected. If you want to compile your own District of Columbia bird list, remember that the District limits include the waters of the Potomac up to the high tide line on the Virginia shore from a little above Chain Bridge to just below Wilson Bridge (though the line there angles sharply north toward the Maryland shore). Columbia Island, including the marina above Rochambeau Bridge, is also part of the District.

Most of the areas described here are parkland, unlikely to be developed in the foreseeable future. They are in parts of the city that are usually safe in daylight hours. They can nearly all be reached by public transportation combined with a short walk. Nonetheless, a car is a real convenience, particularly in Rock Creek Park, the National Arboretum, and East Potomac Park; it is virtually mandatory for the Anacostia Naval Station.

This chapter is far more detailed than the other chapters in this book. It was written almost entirely by David Czaplak, who has a more thorough knowledge of the places to find birds in the District of Columbia than anyone in living memory.

NORTHWEST

1. **Roosevelt Island** is a trap for migrant land birds. The paths south from the memorial follow the ridge of the island and are especially good for thrushes and warblers. As many as twenty-five species of warblers have been found on a May day here. At the south end of the island the path turns east on a footbridge across the marsh. The marsh attracts Swamp Sparrows, Marsh Wrens, Common Snipe, both yellowlegs, and occasionally other shorebirds in migration. American Bitterns, Soras, and Virginia Rails are rarely seen migrants. In late summer Great and Snowy egrets and Little Blue Herons are sometimes seen. By following the path north along the east side of the marsh you can find other vantage points. The path then skirts a wooded swamp attractive to migrant waterthrushes and, in early spring and late

fall, large flocks of Rusty Blackbirds. Side paths branch east here to views of the Potomac River and the Kennedy Center. Least Terns nested on the roof of the Kennedy Center in 1987–88 and could be seen fishing in the river in June and July. The main path continues to the north end of the island and then loops back to the memorial. The entire walk is about 1.5 miles long.

The island has relatively few breeding species. Barred Owls are regular, and Great Horned and Barn owls and Eastern Screech-Owls have all nested in the past. Wood Ducks and, very rarely, Prothonotary Warblers are found in the swamp. Red-winged Blackbirds and Green-backed Herons are in the marsh. Black-crowned Night-Herons from the nesting colony at the National Zoo often roost at the marsh edge or can be seen at dusk, flying over the island.

From the parking lot a bicycle path goes south along the Virginia shore of the Potomac to give views of the tidal mud flats at the south end of Roosevelt Island. Gulls congregate here year-round. Bonaparte's Gulls are common in April, as are Caspian and Forster's terns in summer. Ospreys are likely in spring and fall, Bald Eagles in early winter. Diving ducks are occasionally present.

The entrance to the parking lot for Roosevelt Island is reached only by the *northbound* lanes of the George Washington Memorial Parkway. On leaving the lot you must proceed north along the parkway. If you wish to go south, you can do so by taking the left-hand exit for Spout Run, after about one mile. After another half-mile there is a left-hand lane to make a U-turn for the southbound lane of Spout Run, which leads back to the southbound lanes of the George Washington Parkway.

By public transport: From Rosslyn station exit left on North Moore Street and walk 1.5 blocks to end of street. Turn right, go one block, crossing to far side of North Lynn Street. Turn left, cross I-66 entrance and exit roads, and turn right into maintenance yard. You will see the sign for access to Roosevelt Island.

2. **Fletcher's Boathouse** is on the Chesapeake and Ohio (C&O) Canal about two miles north of Georgetown. The entrance to the parking lot, across from the end of Reservoir Road, is a diagonal left turn as you go north on Canal Road. (Canal Road is one way inbound [south] in the morning rush hour, one way outbound [north] in the evening rush hour.) The picnic area at the river's edge is the best place in the city to see breeding Orchard and Northern orioles and Warbling Vireos. A few pairs of Black and Turkey vultures nest along the river here. In migration this is a concentration point for swallows, Chimney Swifts, gulls, and

Caspian Terns. Migrant ducks are not as plentiful as farther downriver, but diving ducks, such as Bufflehead, Common Gold-eneye, and Greater and Lesser scaup, are often seen.

Walk north from Fletcher's along either the canal towpath or the abandoned railroad. They rejoin after about a half-mile. The woods here have breeding Barred Owls and Wood Ducks. This is the only regular breeding site in the District of Columbia for Prothonotary Warblers. Yellow-throated and Cerulean warblers are rare breeders, not present every year. One mile north of Fletcher's the towpath passes under Chain Bridge. A path leads left under the bridge to a series of overgrown ponds. It is best to climb up via the bike ramp and walk part way out onto the bridge. Here you have an excellent view of all the ponds and the Potomac River. Look for Ospreys, teal, snipe, and American Bitterns in migration and sparrows in winter. Green-backed Her-ons and Canada Geese nest here. Black-crowned Night-Herons are frequent along the water's edge, and Yellow-crowned Night-Herons are rare summer visitors. The bridge is a good vantage point from which to watch whatever is moving along the river: swallows, gulls, ducks, or hawks. Looking north you can see a concrete platform along the east bank. This platform, which is reached by a road from the towpath, approximately marks the District of Columbia/Maryland line. A 1.5-mile walk up the tow-path takes you to Little Falls, another vantage point for ducks.

By public transport: From Dupont Circle station take D4 or D8 bus west to far end of Georgetown Reservoir on MacArthur Boulevard. Cross boulevard and bear left a quarter-mile downhill (northwest) on Reservoir Road to Fletcher's Boathouse.

3. **Battery Kemble Park,** a narrow strip of woodlands and mead-ows, is one of the most pleasant parks in the city and one of the best for migrant land birds. One entrance is on Chain Bridge Road, 0.25 mile south of Loughboro Road. Turn left and follow the dirt road to the parking lot at the bottom of the hill. Walk up to the top of the ring of hills around the lot and check each grove of trees and clump of bushes for migrants. They may be alive with warblers in early morning when the sun is hitting the hilltops. The open areas on the hillside are allowed to grow into meadows each fall; they attract large flocks of migrant Chipping, Field, and other sparrows and once produced a Dickcissel. Lesser num-bers of sparrows winter here. Eastern Bluebirds have occasion-ally attempted to nest here, one of their few breeding sites in the city. In fall and winter, check the open stands of Virginia pines for kinglets, Red-breasted Nuthatches, Pine Siskins, and perhaps crossbills. The dead snags seem to get more than their share of

Red-headed Woodpeckers in fall. South of the parking lot, paths lead along a stream valley and a western ridge to MacArthur Boulevard in 0.5 mile. The forest along the stream is excellent for migrant thrushes and has breeding Veeries. Winter Wrens are common along the overhanging banks of the stream in fall and winter. In the heat of the day, migrants often bathe in the shady pools of the stream.

By public transport: From Tenleytown station take M4 bus west on Nebraska Avenue to Foxhall Road. Junction is at northwest corner of park, with a trail downhill.

4. **Glover-Archbold Park** is another good place to observe migrant land birds, especially warblers and thrushes, but the tall trees make viewing more difficult. The best section is the one-mile length of trail that goes north from Reservoir Road, just east of 44th Street. Veeries breed here, and this is a good place to listen to the evening chorus in June. Rarely, Yellow-throated Vireos and Kentucky Warblers attempt to breed, and an Eastern Screech-Owl is sometimes resident. In winter look for Winter Wrens along the stream and for large flocks of White-throated Sparrows that are concentrated in the brushier west side of the park, away from the trail. Pileated Woodpeckers are common residents.

By public transport: From Dupont Circle station take D4 or D8 bus to first stop west of Georgetown University Hospital on Reservoir Road.

5. **Georgetown Reservoir,** on MacArthur Boulevard just south of Reservoir Road, may attract migrant or lingering waterfowl at any time of year, but October–December and March–April are the best seasons. Viewing light is best before 10 A.M. from the sidewalks on MacArthur Boulevard. You can see the entire reservoir with a spotting scope. Ring-necked Ducks peak in November at nearly one thousand birds, with lesser numbers of Canvasbacks, Ruddy Ducks, Buffleheads, Greater and Lesser scaup, American Wigeon, American Coots, and Pied-billed Grebes. Less frequent, and tending to stay for short periods only, are Oldsquaws, Redheads, Gadwalls, all three species of merganser, Common Goldeneyes, White-winged Scoters, Horned Grebes, and Common Loons. Varying numbers of gulls loaf on the dikes, and there is often a Lesser Black-backed Gull and more rarely an Iceland Gull here in winter. Black and Turkey vultures are infrequent visitors. Once in a while a yellowlegs or a Solitary Sandpiper is seen walking on the bank. The reeds on the far bank are worth watching for American Bitterns. Barn and Rough-winged swallows nest along the dikes.

The reservoir is something of a concentration point for raptors in spring and fall. The flight direction is southwest in fall, northeast in spring, or sometimes along the river. Flights are unpredictable and do not seem to correlate well with weather or flights elsewhere. The best bet is a warm day in early October with northwest winds. A good day might produce a few hundred birds, mostly Broad-winged and Sharp-shinned hawks, but a few dozen are a more likely total. Golden Eagles have been seen twice in October, and a Rough-legged Hawk once in November.

The best vantage point for hawks is the large lawn on the south end of the reservoir. By walking west across the lawn you reach a grassy lane that is an abandoned trolley route. It goes northwest along the far side of the reservoir (but below the dike level, so that you have no views of the water) and is a good place to look for sparrows. Field-loving species such as Grasshopper and Savannah sparrows have been seen along the grassy bank. Yellow-breasted Chats sometimes breed. Hawks pass over the lane, but the view is restricted. In 0.75 mile you reach Reservoir Road and Fletcher's Boathouse.

By public transport: From Dupont Circle Station take D4 or D8 bus to reservoir on MacArthur Boulevard.

6. **Reno Reservoir,** at the intersection of Fessenden Street and 39th Street, is a covered reservoir and has no water for ducks. Instead, the top of the hill is a large fenced-in field that attracts open-country birds in migration (and provides a panoramic view to the west). Park on any of the side streets and climb steeply up to the top of the hill. Walk around the fenced perimeter, peering in at every opportunity. If mowing has been neglected and the grass is six inches or so in height, the field is good for sparrows, such as Savannah, Field, Chipping, White-crowned, and Grasshopper, as well as Killdeer, Eastern Meadowlark, and Cattle Egret. In early May this is the best place in the city to see Bobolinks. In August and September this is also a great place to watch the evening flights of Common Nighthawks, which sometimes pass very low over the hilltop. After a rain the ditches might shelter a Common Snipe or a Solitary Sandpiper.

By public transport: From Tenleytown station, walk north two blocks on Fort Drive to Chesapeake Street, the southern edge of the reservoir grounds.

7. **Rock Creek Park** is the largest block of forest in Washington; it offers a variety of edge, field, and stream habitats. Like all the city's parks, it is best in migration. The high ridge that borders the west bank of Rock Creek between Broad Branch and Military roads is the best warbler trap in the city. The combination of the

ridge of forest, emerging from miles of surrounding city, and the location on the fall line serves to concentrate migrant land birds in great numbers. On some mornings the ridge is swarming with birds. Activity is greatest very early in the day as the sun hits the trees. Later the birds disperse to lower parts of the park.

The park has lost many of its breeding species because of fragmentation and heavy disturbance, but it is still worth a visit in the summer. A few pairs each of Red-shouldered and Broad-winged hawks and Barred and Great Horned owls nest each year. Pileated Woodpeckers are hard to miss. There are more than a dozen pairs of Eastern Screech-Owls, mainly along the streams, but they are rarely seen by day. Looking for them at night is not recommended. Common songbirds include Acadian Flycatcher, Eastern Wood-Pewee, Wood Thrush, Veery, Oven-bird, Scarlet Tanager, and, along streams, Louisiana Water-thrush. Other species that occasionally breed, or attempt to do so, include Brown Creeper and Hooded, Kentucky, Worm-eating, and Cerulean warblers.

In winter the park provides slower birding, with the usual mixed flocks of woodpeckers, chickadees, and kinglets. Accipiters winter in the park and in surrounding backyards. The scattered stands of Virginia pines might have Red-breasted Nuthatches and lingering warblers. The bird feeders at the Nature Center attract Evening Grosbeaks in an irruption year.

Recently the National Park Service has created additional habitat in the park by letting small strips of lawn areas grow up in the fall. These areas are tiny but should not be overlooked—anything might show up. In October they attract hundreds of migrant sparrows. They are the best places in fall to see Orange-crowned, Mourning, and Connecticut warblers.

A suggested route for birders in migration time begins at the Nature Center, just south of the intersection of Military Road and Glover Road. Park in the lot there and check the trees around the edge. Walk into the forest on the short loop trail that begins just behind the Nature Center.

Then walk or drive south on Glover Road. In a few hundred yards turn left into the Headquarters parking lot. The small group of trees in the center are often packed with warblers. At the parking lot entrance walk up the grass bank on the left toward the stables, keeping the corral on your left. Follow the horse trail into the woods, keeping right at the trail junctions. In about a quarter-mile you will see on the right an open field that sometimes contains equipment or haystacks. This is currently the best field habitat in the park and should be worked very thoroughly, especially in October, when Lincoln's Sparrows and

Orange-crowned Warblers are likely. When you leave the field you can explore the network of trails through the forest.

On leaving the parking lot continue south on the main road, which changes its name to Ridge Road. Watch carefully for migrating flocks in the trees. The best spot in the whole park is reached as you emerge into an open lawn just north of the intersection with Ross Drive. At dawn you can watch the migrants drop into the treetops there. Later in the day this spot is also good for hawks. Even loons, cormorants, and swans are frequently seen here. In a finch irruption year, Pine Siskins, Purple Finches, and Evening Grosbeaks pass overhead in large numbers.

Passing the entrance to Ross Drive, you reach an even larger lawn with a corral. This area can be just as good. From the north side, just across the road, a trail leads down a short distance to Rock Creek. A family of Louisiana Waterthrushes is present here each summer.

Continuing south, Ridge Road soon drops down to Broad Branch Road. Turn left on Broad Branch and almost immediately right into the next parking lot. Wood Ducks and Belted Kingfishers nest along the creek. The bushes and goldenrod are attractive to migrants later in the morning, when the sun has warmed the valley. Walk down the bicycle path to Pierce Mill and check the thicket behind the Art Barn for migrating woodcock.

Leaving the parking lot, turn right, and right again onto Beach Drive. Follow it south, watching carefully for the sign for Porter Street, a right turn. About a hundred yards up Porter, turn right again onto Williamsburg Lane. Follow this road up the steep hill, where the road becomes gravel, and park in the dirt lot at the top. You are in Melvin Hazen Park, another good migrant trap. Olive-sided Flycatcher is sometimes seen in the snags around the stone Klingle House, which is park property. A pair of Broad-winged Hawks frequently nest in the woods in front of the house, and House Wrens nest around the barn. Explore the woods and meadows around the house. Singing Alder Flycatchers and Mourning and Connecticut warblers have been found here. From the parking lot trails lead north to Pierce Mill and west to Connecticut Avenue. The latter trail, up a shaded ravine, is favored by migrating Black-throated Blue Warblers and resident Veeries.

This route covers a small portion of the park. North of Military Road it is wider, so migrants are more scattered, but the larger blocks of forest are better for breeding species. The trails offer many days of exploration.

By public transport: From Friendship Heights station take E2 or E4 bus east on Military Road to Glover Road. Walk up the bicycle path that parallels Glover Road to Rock Creek Nature Center.

8. **National Zoo.** From the entrance on Connecticut Avenue, just north of the National Zoo Metro stop, it is a short walk to the Bird House and the waterfowl ponds. The trees around the pond and especially those that overhang the eagle cage support the decades-old Black-crowned Night-Heron colony, about fifty pairs strong in recent years. Nesting begins in late March and continues through July and August. Some nests are placed on the mesh roof of the Outdoor Flight Cage. As the young leave the nests they gather in the duck ponds below; they are very tame. In the evening they make training flights over the nearby apartment buildings and down Rock Creek to feed along the Potomac. Sometimes a few birds winter at the zoo, roosting in the pines south of the flight cage. A juvenile Yellow-crowned Night-Heron was found here one August; they should be watched for.

On a quiet winter day, amid the throngs of House Sparrows and European Starlings you will discover many other sparrow species, attracted by all the birdseed in the outdoor cages and by the heated drinking water. Song, White-throated, and Field sparrows are the most numerous, but even Fox and Tree sparrows have been seen feeding inside the cages. After a hard freeze the wild Mallard contingent at the heated ponds may grow to nearly a thousand. A few dozen wild Wood Ducks also use the ponds, and the occasional American Black Duck, American Wigeon, or Gadwall drops in, but they are hard to tell from the captives unless they fly away. Vultures scavenge in the mammal compounds, and a few Red-tailed, Cooper's, and Sharp-shinned hawks hang about. The route downhill to the seals takes you past some bird feeders (sometimes empty). The ornamental plantings of holly, cedar, and hemlock scattered about the zoo may hold a Barn or Saw-whet owl, but only rarely. Near the polar bear cage you can cross Rock Creek on a bridge and follow the bicycle path through some bits of field and woods that are also worth exploring in winter.

By public transport: Woodley Park–Zoo station.

9. **Dumbarton Oaks Park** is convenient to Georgetown and a good place to look for migrants. In spring the tall trees around the tennis courts and lawns can be splendid for warblers. From R Street just east of 31st Street, walk north down an unmarked lane at the end of the brick wall. Check the dense growth at the bottom, by the stream. In midmorning, migrating warblers often come here to bathe. A right turn before crossing the stream will take you to Rock Creek in a few hundred yards. Turning left

instead, you enter an area of mixed lawns and hedgerows that are especially good in late fall and winter. Work west for about 0.3 mile, checking all the likely-looking thickets. There are two hemlock groves to check for Pine Siskins and other finches. The grove to the north often has a Great Horned or Barred owl or an accipiter. The stream may have a Winter Wren or two. Near the west end of the park is a shady area of hemlocks and rhododendrons, good for migrating thrushes and woodcock. At the west end, a path leads steeply up to Whitehaven Street and Wisconsin Avenue. In winter, watch anywhere in the park for very large flocks of American Robins and Cedar Waxwings that use the park and adjacent private estates. (The grounds of the estate of Dumbarton Oaks, at 31st and R streets, are also good for birds, but visiting hours are restricted.)

By public transport: From Dupont Circle, take D2, D4, D6, or D8 bus west on Q Street to 30th Street. Walk uphill (north) on 30th Street to R Street and turn left. Take the lane north downhill at the east end of the brick wall.

NORTHEAST

10. **The National Arboretum** is attractive to a variety of hawks and sparrows in winter, and the extensive conifer groves offer the best owl hunting in the city. The main entrance is at the end of R Street, reached by turning east on R from Bladensburg Road, a half-mile south of the intersection of Bladensburg and New York Avenue, U.S. 50. The gate is open on weekdays at 8 A.M. and on weekends and holidays at 10 A.M. Get a map at the gate and drive east about one mile to Hickey Hill, which is covered with groves of cedar, spruce, and white pine. Barred and Great Horned owls are resident. One or more Saw-whet Owls are present at least every other winter but might take hours to find. They are most likely to be here from late December to February in the spruce grove on Hickey Hill or in the vast ornamental conifer gardens to the northwest. Yet they might turn up anywhere, in a holly or vine tangle. Less common but to be hoped for are Long-eared and Barn owls.

The conifer groves also have large kinglet flocks and Red-breasted Nuthatches in a good winter. The Virginia pines are especially attractive to lingering warblers, which are sometimes found in December and January by the diligent searcher. They associate with the kinglet flocks, but they tend to stay near the tops of the pines and out of sight. One memorable day in late December produced an Orange-crowned Warbler, a

Black-throated Blue Warbler, and a Northern Oriole. Other species found in northeast Washington pine groves in winter include Cape May, Blackburnian, Pine, and Palm warblers. The hemlock groves are good for Pine Siskins and, rarely, White-winged Crossbills. Red Crossbills, almost as rare, seem to prefer the pines. Neither crossbill is present in most winters.

South of Hickey Hill is a large open area with scattered hollies, hemlocks, and fruit trees. The largest robin and waxwing flocks in the District of Columbia are usually found here. Watch these flocks carefully; a Bohemian Waxwing was amazingly present one winter.

East of Hickey Hill is the Anacostia River, and there are vantage points from which to watch in winter for diving ducks, especially mergansers, and the occasional Bald Eagle. Several paths lead down to the bank. (It is more productive to look for water birds on the Potomac, however.) Other parts of the Arboretum are worth exploring in winter. This is one of the few breeding sites in the city for Bobwhite.

By public transport: From Stadium-Armory station, take B2 or B4 bus northeast up Bladensburg Road to R Street and walk east to the arboretum gate.

SOUTHWEST

11. **Constitution Gardens** at the west end of the Mall is the ideal migrant trap: an island of greenery surrounded by pavement and water with few places for the birds to hide. On some days every tree and bush shelters tired thrushes, warblers, and sparrows. On other days the gardens are empty. The best times are mid-April to late May and September to October, after south winds in spring or passage of a cold front in fall.

The gardens are on the Mall south of Constitution Avenue and west of 17th Street. On a weekend morning park on Constitution or Virginia avenue. (On a weekday, parking is very difficult.) At the large pond you will find a gathering of Ring-billed Gulls, Mallards, and, in migration, American Wigeons. Other migrating waterfowl seen here include Redhead, both scaups, all three mergansers, and Northern Shoveler, as well as Pied-billed Grebe. (The pond is frozen in very cold weather.) The lawn areas attract large flocks of Chipping Sparrows, and such urban rarities as Vesper, Grasshopper, and Clay-colored sparrows. Walk east, then south around the Reflecting Pool, to its south side. The grove of trees around the war memorial can be especially good, and the azalea bushes hold large flocks of sparrows.

The Reflecting Pool may also attract ducks and once, in March, had a flock of American Pipits walking on the stone edge. Continue west, then north past the Lincoln Memorial to return to your starting point. This loop walk is about one mile long. If it is a good day for migrants, check the grounds of the Pan American Union at the northwest corner of Constitution Avenue and 17th Street. These gardens can be just as rewarding.

By public transport: From Farragut West station walk south on 17th or 18th street eight blocks to the northeast corner of Constitution Gardens.

12. **The Tidal Basin and East Potomac Park** (known unofficially as Hains Point) are the best areas for gulls in winter and, under appropriate conditions, ideal for loons, grebes, ducks, and shorebirds. Gulls are present all year, but the buildup begins in August with Ring-billed and Laughing gulls. Herring and Great Black-backed gulls begin arriving in fall. The best period for Glaucous, Iceland, and Thayer's gulls is from January to March. One or more Lesser Black-backed Gulls are present from late September to early April, and you have at least a 50-50 chance of seeing one from October to March on any given day. Franklin's Gull turned up in four of the six years 1984–89, in May, June, or September.

Typically about 1,000 gulls are present in the daytime. Many more come in to roost just before sunset, usually going to the Tidal Basin, but sometimes using the golf course on Hains Point or the Potomac. They leave before dawn, so it is best to visit in late afternoon. If there is a week or more of subfreezing temperatures, Washington Channel and the Potomac freeze over. This condition usually results in a fish kill, and gull numbers skyrocket. On one such day in late January, the channel was packed with 20,000 gulls, including one Iceland, four Lesser Black-backed, and four Glaucous gulls. If the freeze continues, the flocks shift down the Potomac to stay near open water. (Viewing points in Alexandria are described in the next chapter.) In spring and fall the gulls prefer the golf course, and Lesser Black-backed and Franklin's gulls are more likely to be found there than on the river. In April migrating Bonaparte's Gulls pass the point by the hundreds, and in summer and fall Caspian and Forster's terns fish the river.

East Potomac Park attracts a large number of interesting birds in migration, but weather is the critical factor. Strong northeast or southeast winds with heavy rain force down many ducks and shorebirds that might otherwise pass over without stopping. During sunny, pleasant weather with south or northwest winds, birds are less inclined to stop and more likely to be driven off by the golfers and tourists. As a rule, the worse the weather the better your chances of finding a rarity. Large numbers of Common Loons and Horned Grebes stop under such conditions, especially in November and March. Search among them for such local rarities as Red-throated Loon, Red-necked Grebe, Eared Grebe, Brant, Oldsquaw, and all three scoters. The golf course attracts shorebirds, especially if rainpools form and last for a few days. The pools are best checked at dawn, before the mowing crews are out. From April to June and August to October watch for Black-bellied Plovers, both species of yellowlegs, "peeps," Pectoral Sandpipers, and Common Snipe. Rarities that have been found include Lesser Golden-Plover, Upland Sandpiper, and even Ruddy Turnstone. In spring Cattle Egrets are seen regularly and Glossy Ibis occasionally. Huge flocks of swallows are present on some days in migration.

In winter, a Merlin or accipiter usually winters, and Bald Eagles might be seen in the trees along the river. Short-eared Owls rarely roost on the golf course, and Lapland Longspurs, Horned Larks, and American Pipits are occasionally seen, especially on bare areas.

To bird the park, start at the intersection of 15th Street (Southwest) and Ohio Drive, just east of the Jefferson Memorial. Turn south on Ohio Drive and follow it past the golf course clubhouse. The drive becomes one-way and goes clockwise around the point. Stop frequently to scan Washington Channel and the golf course. At the point check all the trees and bushes for migrating passerines and scope the Anacostia and Potomac rivers for waterfowl. Many birds come and go, so it is sometimes worth waiting there. Continue north along the Potomac, watching for flooded ditches and rainpools. After going under the bridges at the north end, turn left over the small bridge at the mouth of the Tidal Basin. Park here and check the basin, the playing fields, and the holly grove (for waxwings and robins in winter). From here it is a short walk across Independence Avenue to Constitution Gardens. Take note that Ohio Drive is often closed to auto traffic on weekend mornings for footraces, usually announced in the Friday "Weekend" section of the *Washington Post.* It is also usually closed to cars on weekend afternoons except in winter; you do not want to visit the park then anyway, because of the huge crowds.

By public transport: From Smithsonian station, walk west on Independence Avenue to Raoul Wallenberg Place (the extension of 15th Street) and turn left. Cross to the Tidal Basin and circle it clockwise. After crossing a small bridge turn left at the

"East Potomac Park" sign and walk under the overpass to Hains Point.

13. **Anacostia Naval Air Station** (and Bolling Field) formerly attracted shorebirds in migration and Short-eared Owls in winter, but most of the habitat has been destroyed. It still attracts some shorebirds, especially snipe, an occasional Sora, and, once, a Seaside Sparrow.

The entrance is at South Capitol Street and Firth Sterling Avenue, most easily reached from downtown Washington by crossing the Anacostia River on the South Capitol Street Bridge and taking the first exit ramp on the right. The guards normally let birders in, but all road signs must be carefully observed. On entering, you are in a large lawn area marked by a yellow airplane on permanent display. Park in one of the roadside lots and, if you are here early in the morning, check for Lesser Golden-Plover and possibly Buff-breasted Sandpiper. Near the plane is a ditch, usually overgrown and often full of rainwater. Tramp through the ditch to find Snipe, Soras, and Savannah Sparrows. Walk south from the plane on the road that passes to the east of some enclosed baseball fields. In about 200 yards you will find a large bed of cattails, usually with some open water in the center. Check it for bitterns, rails, and shorebirds. To minimize stress on the guards on duty, do *not* look at the presidential helicopters, which are sitting some distance away behind a fence.

After extremely heavy rains these lawns are flooded, and gulls, terns, and shorebirds may be found. Rarities include American Avocet, Hudsonian Godwit, Baird's Sandpiper, Red-necked Phalarope, and Sooty Tern.

By public transport: Take a taxi.

For more Washington birding, read on: you can look into the District of Columbia from Virginia.

From National Airport to Mount Vernon 5

Among the most thoroughly explored birding areas around Washington is the stretch between the Potomac River and Shirley Highway, I-395, which becomes I-95 south of the Capital Beltway (I-95/495). Despite its urban character, Alexandria contains a diversity of habitats that make for interesting birding all year long, and the number of

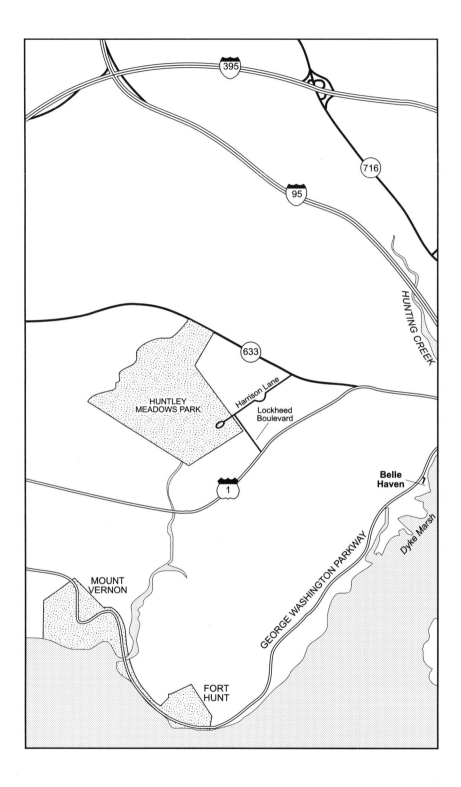

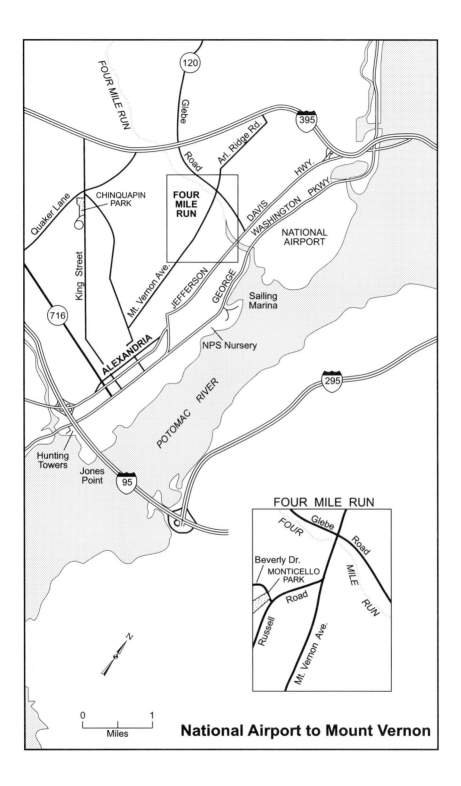

National Airport to Mount Vernon

FOUR MILE RUN

120

Glebe

Arl. Ridge Rd.

395

Road

DAVIS

HWY.

WASHINGTON PKWY.

Quaker Lane

CHINQUAPIN
PARK

**FOUR
MILE
RUN**

NATIONAL
AIRPORT

King Street

Mt. Vernon Ave.

JEFFERSON

GEORGE

Sailing
Marina

716

NPS Nursery

ALEXANDRIA

POTOMAC   RIVER

295

Hunting
Towers

Jones
Point

95

N

FOUR MILE RUN

FOUR

Glebe

Road

Beverly Dr.
MONTICELLO
PARK

Road

MILE

RUN

Russell

Mt. Vernon Ave.

0   1
Miles

parks and refuges downriver are expanding to provide increasing access to natural areas. The river south to Wilson Bridge, up to the high tide line, is within the boundaries of the District of Columbia; below the bridge it is in Maryland except for well-defined bays. Those who keep separate lists will need to study jurisdictional boundaries and keep the birds they see on and over the Potomac separate from those they see in Virginia.

At the mouth of **Four-Mile Run** is a bay on the Virginia shore of the Potomac, immediately south of National Airport, where there are extensive tidal flats. All the flats and the bay are in the District of Columbia (and are shown as #14 on the District map at the end of the book). A visit within an hour of low tide is most productive, and a scope is essential.

For best viewing in early morning light, enter National Airport and follow the signs for Satellite Parking B and C. Drive to the north edge of the bay, where you can scan the flats to the south for gulls and terns. Shorebirds prefer the flats to the west, near the boathouse. (The parking fee is minimal or may be waived if you are there for just a short time.)

Later in the day drive south on the George Washington Memorial Parkway from the airport. At 0.75 mile after crossing the bridge over Four Mile Run, turn left into Daingerfield Sailing Marina. Park and walk to the shoreline to look north over the flats. A bicycle trail leads north to the stream and the boathouse by the best shorebird area. Climb down the bank and peer through the bushes to get a better view, trying to avoid flushing the birds.

District rarities found here recently include Parasitic Jaeger, Piping Plover, Willet, Sanderling, Baird's and White-rumped sandpipers, and white-winged gulls. Diving ducks and teal winter here, and egrets and Little Blue Herons may be here in late summer.

The **old National Park Service nursery** runs along the river south of the Daingerfield Sailing Marina. It is good for migrant passerines and wintering hawks and sparrows, and a likely place to look for owls. Park in the lot on the south side at the far end of the marina entrance road. Enter the nursery by way of the gate barring the service road south.

In the 1980s much of the upper tidal Potomac from Alexandria south developed huge mats of hydrilla, partially submerged aquatic vegetation that greatly increased the number of water birds feeding in the area from August to Thanksgiving or later. The mats increase or decrease in size depending on the amount of rainfall, but they are an indication of the clarity of the river water, and the improving control over the sources of its pollution.

The Alexandria waterfront. The hydrilla and the Potomac River between Blue Plains Sewage Treatment Plant and Wilson Bridge currently attract the greatest number and variety of terns and shore-birds in the District of Columbia and are also excellent for ducks and gulls. The best vantage points are in Alexandria (#16 on the District map), and a scope and afternoon light are required.

Begin in north Alexandria at 3rd and Fairfax streets, three blocks east of Washington Street and just south of the VEPCO power plant. Walk to the bluff above the Potomac and check the river for ducks and the pilings for cormorants and terns. Ospreys may nest on the large wooden dock across the river. Walk north along the bicycle path as far as the outlet for the plant to check for ducks in cold weather and gulls and terns at any time. Drive south along the waterfront. The cove at Madison Street is worth checking.

At Oronoco Street there is an elevated wooden deck that offers the best view of the hydrilla mat. At low tide shorebirds rest and feed on it, sometimes in the hundreds. These are mostly yellowlegs, Pectoral Sandpipers, and Dunlins, but Lesser Golden-Plovers, Avocets, Hudsonian Godwits, Wilson's Phalaropes, Sanderlings, dowitchers, and Stilt Sandpipers have been seen. The smaller sandpipers can be identified only with high-powered scopes under optimal viewing conditions. Terns also like the hydrilla, and hundreds of Forster's Terns gather here in late summer and fall. Smaller number of Caspians and up to a dozen Black Terns may be present. Rarities include Common, Royal, and Gull-billed terns and Black Skimmers. Herons and egrets are common, and a Tricolored Heron is occasionally found. Other rarities seen here include frigatebird (species) and Parasitic Jaeger, so it pays to stay alert. In winter Tundra Swans and diving and dabbling ducks feed on the remnants of the hydrilla, and Bald Eagles are regular on the flats or in the trees on the far shore. Great Cormorants may turn up from September to December.

Continue south, turning right at Green Street and left on Royal Street. Just before Royal Street passes under Wilson Bridge, turn left on South Street and follow it to the river. Park and check the hydrilla (here much closer to shore). Watch for Peregrines under the bridge. If you are interested in keeping state lists you will need to have a map showing the District of Columbia/Maryland line running diagonally across the river. Most of the river is in the District, but the southeast corner is in Prince George's County, Maryland.

By walking south from the parking lot under the bridge, you reach **Jones Point,** a park on **Hunting Bay.** The woods on the point are rewarding in migration and the swamp shelters breeding Willow Flycatchers. The bay is in Virginia. An alternative parking area is next to the playing field on the south side of the bridge. If you walk through the trees on the right side of the playing field, there are excellent views west in morning light of the water birds gathered on the hydrilla there. This area can also be reached from South Washington Street.

Just north of the beltway turn east on Green Street and south on South Royal Street. For another view of the cove, walk west from the playing field lot to a short pier next to the east side of Hunting Towers Apartments.

From South Washington Street, most of Hunting Bay and the **mouth of Hunting Creek,** which empties into the bay, are best viewed from the bridge over the creek just south of the point at which the road goes over the beltway, passes Hunting Terrace Apartments and Porto Vecchio, and becomes George Washington Parkway again. Parking possibilities close to the bridge are limited; the best bet is to use the lots at Belle Haven Picnic Area south of the bridge, on the left, and to walk back on the bicycle trail with your scope.

The pilings on both sides of the bridge are renowned for their high tide populations of terns, especially from July to September, and gulls (including occasional Franklin's, Iceland, Lesser Black-backed, and Glaucous) from September to April. Assorted herons turn up along the shoreline, and Double-crested and (rarely) Great cormorants, Greater and Lesser scaup (in winter), and Ruddy Ducks may be farther out in the river. Look for Pintail, Green-winged Teal, and other dabbling ducks from late September on. Shorebirding may be excellent, especially on the flats west of the bridge in September. Lesser Golden-Plovers, Avocets, a Hudsonian Godwit, Curlew and Sharp-tailed sandpipers, Ruffs, and all three phalaropes have turned up (*very* rarely) among the peeps, yellowlegs, and other common species. Check the newspaper for days when the tide is close to low at the time you want to visit, and plan to be there when the sun will not be in your eyes. This outstanding natural area is under serious threat of destruction at this writing; it may be crossed by a new fourteen-lane bridge over the Potomac.

Dyke Marsh Wildlife Preserve lies immediately south of Belle Haven Picnic Area, east of the parkway just below Hunting Creek. After checking for waterfowl and cormorants, including the occasional Great Cormorant on a snag in the Potomac, from the picnic grounds (in season), walk around the woods at the south end. Cross the paved road that goes out to the Belle Haven Marina and follow the trail that leads eventually out to the marsh. Breeding birds have included Common Moorhen, Least Bittern, Willow Flycatcher, Marsh Wren, Yellow Warbler, and Common Yellowthroat. In autumn and winter look for hawks, waterfowl, and sparrows; both spring and fall are good for rails and all kinds of land birds. In the swampy woods west and south of the marsh, you may find Great Horned Owls, Fox Sparrows, and Rusty Blackbirds. The woods are most easily reached from the bicycle path along the parkway, which will take you to a pleasant bridge from which you can inspect another part of the marsh.

The eastern side of **Fort Hunt Park** and the woods around the parking lots for Mount Vernon, on both sides of the parkway, can be good in migration for woodland species—flycatchers, thrushes, warblers, and tanagers. The latter area has resident Great Horned and Barred owls and White-breasted Nuthatches, and nesting warblers include Black-and-white, Kentucky, and Ovenbird.

The Potomac is best birded from south to north—you will be next to the river. As you will be looking east, you will enjoy it most in the afternoon. Look for Bald Eagles and, in winter, sometimes extraordinary numbers of waterfowl, especially bay ducks, along the river. There are several parkway overlooks on the right as you drive north from Mount Vernon to Dyke Marsh. Stop first at Riverside Park, a good stop for Common Mergansers 1.2 miles north of Mount Vernon. Try the pull-off at 1.9 miles, another at 3.3 miles, and Collingwood picnic area at 3.6 miles. A worthwhile stop at the south end of Dyke Marsh is at 5.8 miles, unfortunately on the wrong side of the road.

Away from the river two other spots north of the beltway are recommended:

To visit **Monticello Park,** leave I-395 on Arlington Ridge Road or Glebe Road East. Where these two roads intersect, the former becomes Mount Vernon Avenue and crosses Four Mile Run. Turn right on Russell Road, the third street on the right south of Four Mile Run, running off at an angle from Mount Vernon Avenue. After five short blocks on Russell Road, turn right on Beverly Drive. In 100 yards or so, where Beverly Drive bends sharply right, you will see the park on your left. Monticello Park is a postage-stamp area with tall tulip trees, oaks, and a little stream. It is a favorite spot for warblers in the spring and is at least as rewarding in midmorning as it is at dawn. What is so entrancing for birders is the opportunity to see the birds bathing as close as one's binoculars will focus. The variety of species is extraordinary.

Chinquapin Park is 1.0 mile from I-395; take the exit for King Street East, drive through the major intersection with Braddock Road and Quaker Lane, and continue to T. C. Williams High School on the right. Just beyond the school go right onto Chinquapin Drive, a loop road, and park near the garden plots, which often attract hummingbirds, finches, and sparrows. Bird the edge of the woods on the left for migrants, and then take the trail in the corner behind the gardens. The left branch will lead to a stream and beyond it to meadows and woods of Virginia pine, the latter especially good for Red-breasted Nuthatches, Golden-crowned and Ruby-crowned kinglets, Pine Warblers, Pine Siskins, and other irregular pine-loving visitors, as well as a considerable assortment of migrants.

6 From Huntley Meadows to Mason Neck

The most renowned park in southern Fairfax County is **Huntley Meadows,** an extensive freshwater wetland just southwest of Alexandria. From the Capital Beltway (I-495) take U.S. 1 south 3.2 miles to Lockheed Boulevard. Turn right and go 0.6 mile to Harrison Lane. Turn left into the park. The park itself is open from dawn to dusk, but the visitor center, where a bird list and trail map are available, is open from 9 A.M. to 5 P.M. on weekdays except Tuesdays, and on weekends and most holidays from noon to 5 P.M. (Restrooms are inside the visitor center.) From the parking lot take the path through the mixed hardwood forest to the boardwalk, 0.6 mile long, which extends into the swamp that surrounds a 100-acre beaver pond. The path leading to an observation platform 17 feet above the pond branches off to the left before the boardwalk begins.

Some 150 species have been seen from the platform. Breeding species of special interest include Pied-billed Grebes (no longer regular), American and Least bitterns, Common Moorhens, and King Rails in the wetland; Yellow-crowned Night-Herons, Red-shouldered Hawks, American Woodcocks, Barred and Great Horned owls, Ruby-throated Hummingbirds, Acadian Flycatchers, Brown Creepers, Hooded and Prothonotary warblers in the forest or among the dead and dying trees around the pond, and Northern Bobwhites, Prairie Warblers, and Yellow-breasted Chats in drier brush along the service road through the northern section of the park.

Five species of woodpecker are permanent residents, and Red-headed Woodpeckers are present irregularly. In late summer assorted herons and ibis come to visit, and in autumn Sharp-shinned and Cooper's hawks are among the raptors that can be studied from the platform. In spring and fall Green-winged and Blue-winged teal, Soras and Virginia Rails, both species of yellowlegs and Solitary and Spotted sandpipers, six species of swallows, and twenty species of warblers are among the transient species. Swamp Sparrows and Rusty Blackbirds are common in winter, and Winter Wrens forage around the fallen trees at the start of the boardwalk.

Two other small refuges lie within the U.S. Army reservation of Fort Belvoir but are open to the public. The larger of the two, **Accotink Bay Wildlife Refuge,** is a 700-acre sanctuary. From I-95 take the Fort Belvoir/Newington exit (#56), keep to the right, and follow Backlick Road east 3.4 miles to U.S. 1 at the (unmarked) Tulley Gate, by which you enter the fort. (If you are coming south on U.S. 1, you will pass the main gate to Fort Belvoir on the left, go under two over-passes, and turn left at the first light into Tulley Gate.) In 0.6 mile the parking lot for the refuge is on the right. There is a marsh in the center, with a boardwalk leading out to an observation platform, beaver ponds, and trails through old hardwood forest. A checklist of the 206 species recorded in Fort Belvoir is usually available at the trail head by the parking lot. Arrive at this refuge before 3:30 P.M.; the entrance road becomes one-way west in afternoon rush hour.

The **Jackson Miles Abbott Wetland Refuge** is on the west side of U.S. 1. From Tulley Gate go north 1.6 miles, pass the light at the junction with Va. 235 (which goes to Mount Vernon), and go im-mediately left (within a few feet!) on Old Mill Road. It dead-ends in 0.5 mile at Pole Road; turn right and drive 0.2 mile to the parking area for the refuge. (From Lockheed Boulevard, the access road to Huntley Meadows, Old Mill Road is a right turn 4.4 miles south on U.S. 1.) This sanctuary, established in memory of the birder who recorded the birds of the Alexandria area and Fort Belvoir for over forty years and who created the Fort Belvoir checklist, is less than 150 acres in extent. There is hope that it will eventually be linked with Huntley Meadows. It lies along Dogue Creek and includes swamp, marsh, and an artificial pond. The drowned trees attract woodpeckers and raptors, and dabbling ducks are common winter visitors. The edges are worth checking for migrants.

Pohick Bay Regional Park is the most dependable place close to Washington to see Bald Eagles. Take I-95 south from the junction with I-395 about 6.5 miles to the exit (mile 0.0) for Lorton, Road 642. At the T-junction with Road 642 go left; at 0.8 mile turn right on Road 748, which meets U.S. 1 south in 0.2 mile. In another 0.9 mile turn left on Va. 242, Gunston Road. All these turns are very clearly marked for Pohick Bay Park and Gunston Hall. The entrance to the park is on the left in 3.3 miles from U.S. 1.

Drive down to the boat-launching ramp and scan Pohick Bay from there. From the picnic tables at the left end of the paved area look across the water to see the eagles and up the bay to look for ducks. On winter mornings the eagles drift across from their roost on the far side. A nature trail that starts at the parking lot just back from the boat-launching area parallels the river and has some overlooks that enable you to see farther up the bay. It is a good path from which to see woodpeckers, even Red-headeds on occasion, as well as her-ons, waterfowl, and perched eagles not visible from the ramp area. A fee is collected in the busy season except from Northern Virginia residents.

Turn left when you leave this park. At the entrance to Gunston Hall, Va. 242 becomes Road 600 and its surface deteriorates a bit. In 1.1 miles go right at the sign for **Mason Neck Management Area,** a territory combining a national wildlife refuge, a state park, some undeveloped land owned by the regional park authority, and Gunston Hall Plantation. There is a refuge information kiosk on the left 0.7 mile down the road, as well as the head of Woodmarsh Trail, a three-mile loop that can be shortened by either of two cross-trails. It leads through bottomland deciduous woods to the edge of the Great Marsh. Closed from December to March to protect nesting Bald Eagles, it is worth checking for them the rest of the year, as well as for Red-headed Woodpeckers and for Winter Wrens (at the bridges). Look for Barred and Great Horned owls, Woodcock, Green-backed Herons, Wood Ducks, and Kentucky Warblers, as well as the eagles, of course. In migration songbirds may be as close as the trees around the parking area. The marsh is home to Swamp Sparrows and blackbirds, as well as both species of bittern, herons and egrets, harriers in winter, and Ospreys in summer.

In another 0.4 mile, you will come to the entrance tollbooth for Mason Neck State Park. The booth is staffed from 9 A.M. onward between Memorial Day weekend and Labor Day, but the park is open from 8 A.M. to dark. In 1.8 miles you will reach the park visitor center, which is closed in winter. You can pick up a park brochure and trail map from a box outside the door. Scan Belmont Bay for waterfowl and eagles, which may be sitting on the ice if the bay is frozen over; the light is best in the morning. Kanes Creek Trail is a good place to look for Hooded Mergansers in fall and Wood Ducks all year. It can also produce views of eagles, but the far end of the trail is closed in winter to protect the eagle roost. The Bay View Trail looks over Belmont Bay, then crosses a tidal creek via a boardwalk and returns through mature woods. It is the best trail in the park for Acadian Flycatchers, Yellow-throated Vireos, Yellow-throated and Prothonotary warblers, and Scarlet Tanagers, and for migrating songbirds.

After (or instead of) visiting the park continue down Road 600 for 1.4 miles to Great Marsh Trail, a 0.75-mile trail through an old mixed hardwood forest. It is open all year. The views over the marsh are spectacular. You will want to take your spotting scope to study the raptors perched on distant snags, the herons in the marsh, or the ducks out in the river. Be alert on all these trails for Red-headed Woodpeckers.

The parks in Arlington County and northern Fairfax County are pleasant enough for hiking in winter, but the birdlife is rarely exciting: plenty of ducks on the river, woodpeckers and chickadees in the woods, perhaps a Barred Owl or even a Wild Turkey and assorted winter finches and sparrows.

In migration, however, the parks come into their own, particularly in the spring, when there are not only songbirds overhead but a splendor of wildflowers underfoot.

The most remote of the parks described here is about 18 miles from Key Bridge and only 8 miles beyond the Capital Beltway, I-495 (though one of the others extends far into Loudoun County).

Glencarlyn Park, along with **Long Branch Nature Center,** which has been carved out of it, lies on the southwest edge of Arlington County. Access is possible from dawn to dusk into Glencarlyn, and, from 8 A.M. on, into the parking lot of Long Branch. You can walk from one into the other as you please.

Approach them both from Washington via Arlington Boulevard, U.S. 50, by taking an exit right for Carlin Springs Road, 1.7 miles west of Glebe Road, Va. 120. At the foot of the exit ramp, turn right at the T-junction. At 0.3 mile a left turn on Fourth Street takes you to another T-junction with Harrison Street (at 0.7 mile), which borders Glencarlyn Park, with entrance roads both left and right, leading down to parking areas at the bottom of the hill.

You will reach the nature center by continuing on Carlin Springs Road past the Northern Virginia Doctors Hospital to the first left turn beyond, which appears to be a driveway for the Medical Center, just 0.7 mile from the foot of the U.S. 50 ramp. Follow the driveway back to the nature center parking lot; check the ponds on the way in. From the path to the nature center take the trail uphill to the left. It curves right, up to an overgrown meadow on the ridgeline and down again behind the center. Scan the scrub in the meadow and trees around the edge. Then take the paved trail along Long Branch, through the Glencarlyn parking area, across the bridge over Four Mile Run to the bicycle trail that runs under the power line. Climb up to the path along the ridge beside the trail and follow it left to its end then right as far as a cluster of tall, dead trees up the hill across the ditch beside you. This very short walk can be so rich in birds that it may take you most of a morning.

The nature center, like the others mentioned in the pages that follow, can provide you with a trail map, a bird list, and good advice

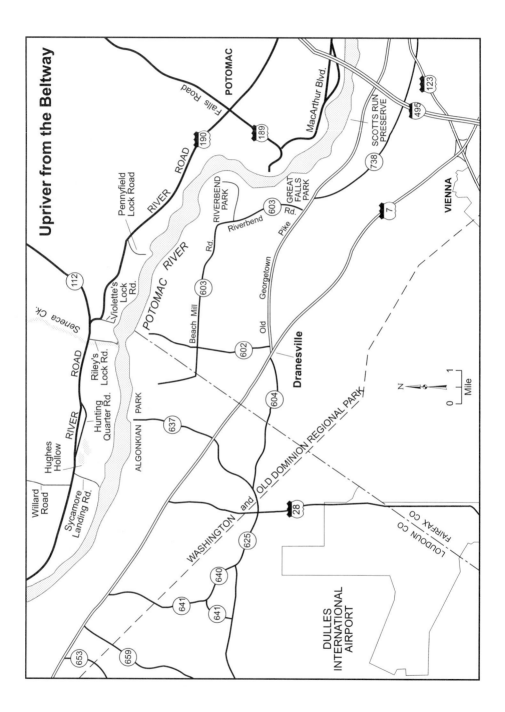

Upriver from the Beltway

POTOMAC

Falls Road

MacArthur Blvd.

SCOTTS RUN PRESERVE

123

495

190

189

738

RIVER ROAD

Pennyfield Lock Road

RIVERBEND PARK

GREAT FALLS PARK

603

Riverbend Rd.

7

VIENNA

112

Violette's Lock Rd.

POTOMAC RIVER

Rd.

603

Georgetown Pike

Seneca Ck.

RIVER ROAD

Riley's Lock Rd.

Beach Mill

602

Old

Dranesville

Hunting Quarter Rd.

ALGONKIAN PARK

637

604

N

0 1 Mile

Hughes Hollow

OLD DOMINION REGIONAL PARK

Willard Road

Sycamore Landing Rd.

WASHINGTON and

28

625

LOUDOUN CO FAIRFAX CO

640

641

641

DULLES INTERNATIONAL AIRPORT

653

659

from the birders on the staff. It is open from 10 A.M. to 5 P.M. Tuesday to Saturday, and from 1 P.M. to 5 P.M. Sunday.

Upton Hill Park, on the south side of Wilson Boulevard off Patrick Henry Drive, just east of the Arlington/Fairfax county line, is a favorite place on evenings in late August to see Common Nighthawk migration and to watch Chimney Swifts swirling into nearby chimneys. As the highest point in the vicinity, well above the local rooftops, it is a fine place to watch hawks (when the winds are from the northwest) from mid-September to November. Bring something to sit on; beach chairs are highly recommended. The woodland trails can produce an interesting assortment of songbird migrants, and the warbler list is a long one. There are many other recreational options in this park besides birding, and you may sometimes find it swarming with people.

Scotts Run Nature Preserve (formerly Dranesville District Park) is just outside the beltway on Va. 193, Georgetown Pike. Turn in at the second entrance 0.7 mile downhill from the beltway exit for Great Falls, and park in the lot beside Scotts Run.

This is a Fairfax County Park with no facilities or visible staff or official bird list. Breeding and winter bird surveys have been conducted here since 1971.

The census plot is in quite uniform upland hickory-oak woods and the record shows that only the most common deciduous forest birds breed and winter regularly in this park. The interesting area in migration is along Scotts Run, where there are dense thickets, tall trees overhanging a stream, steep slopes, and, down by the Potomac, a rocky gorge lined with hemlocks. If you feel like climbing to the heights above the gorge, keep to the trails close to the river for a maximum of birdlife.

It would be a pity to ignore the flowers, though, especially since there is a fine little pamphlet called *A Habitat Guide to Spring Wildflowers of Dranesville District Park.* You can buy it for a dollar at the Interpretive Center at Riverbend Park (see below). The booklet provides you with a map of the park showing the trails and the six habitats, a species list for each habitat, and an overall checklist indicating blooming time and abundance. It is invaluable for all the parks in the area.

Great Falls Park, a national park, is 4.3 miles up Georgetown Pike from I-495; turn right at the light onto the entrance road. It is the only park with a year-round entrance fee, which you can avoid by arriving before 8 A.M., though you will also miss receiving a handy little map of all the trails.

Make a U-turn to the right just beyond the tollbooth and park at the end of the road. In migration, birds are likely to be in all the trees along the edges of the lawns, roads, and parking lots, but the most productive areas are the Swamp Trail and the woodland behind the rest rooms, close to the old Potowmack Canal and the river. Cerulean and Kentucky Warblers are among the nesting species. Look for Acadian Flycatchers and Yellow-throated and White-eyed vireos.

It is worthwhile to follow the service road that goes toward the south end of the park and to loop west from it on the Swamp Trail.

North of the visitor center take the trail up to the dam above the falls. From November to April Ring-necked Ducks are regular; Buffleheads, Common Goldeneyes, and Common and Hooded mergansers are often present, and are sometimes joined by other species. You may find loons, grebes, and gulls above the dam. This end of the park is good for Warbling Vireos and Prothonotary and Yellow-throated warblers.

Throughout the park look for Brown Creepers, Winter Wrens, and both kinglets in winter, Pileated Woodpeckers and Eastern Bluebirds all year.

Riverbend Park, another Fairfax County Park, is only a mile upstream from the dam along the river trail, but it is 5 miles by road. Go back to Va. 193 and turn right (mile 0.0). At 0.3 mile turn right again on Riverbend Road and follow the signs to the park, forking right on Jeffery Road at 2.3 miles. After 3.2 miles the entrance sign gives you a choice: the road straight ahead goes to the Interpretive Center, while a right turn takes you to a visitor center, rental boats, a boat ramp, and fine views over the river. A fee is charged on the latter road on summer weekends except to Fairfax County residents. Look for Ring-billed and Bonaparte's gulls and Common Mergansers in March and April; later in the spring it is a splendid vantage point from which to watch migrating swallows, swifts, and nighthawks.

You can join the trail along the river here, but you may do better going back to the road to the Interpretive Center, open from 9 A.M. to 5 P.M. on weekdays and from 1 P.M. to 5 P.M. on weekends, and closed altogether in January and February. The park grounds are open from dawn to dusk. No fee is charged to visitors who enter here, and you can make a circuit down to the boat ramp, coming back along the riverside trail that goes upstream.

There is much variety of habitat in this park, including overgrown fields and brush; it is the best of the six in which to look for Wild Turkeys and sparrows. You can assume that the birds you may see are very similar to those elsewhere along the river.

The **Washington and Old Dominion Railroad Regional Park,** known for short as the W&OD Trail, is 100 feet wide and nearly 45 miles long, most of it lying south of Va. 7 and roughly parallel to it. This old railroad bed extends from I-395 in Shirlington in Arlington County to Purcellville in Loudoun County. A trail guide for the entire route is available.

Much of it goes through residential districts and is heavily used as a bicycle and jogging trail and a neighborhood footpath, but west of Vienna there are fine stretches for birds, wildflowers, and scenery. In Fairfax County, try the two miles from Vienna to Difficult Run. The trail runs beside Piney Branch and crosses it several times, passing through bottomland forest and open fields. You can park in Vienna at the library on the corner of Va. 123 (Maple Avenue) and Center Street. Farther west, sample the trail from Ashburn to Tuscarora Creek, some five miles of coniferous and deciduous woods, orchards, pastures, ponds, and creeks. For points of access, east to west, take Va. 7 into Loudoun County and turn left (south) on Road 641, Ashburn Road, or on Road 659 or on Road 653. Each crosses the trail in two miles or less.

Among the breeding species to look for are Black Vulture, Willow Flycatcher, Yellow-breasted Chat, Summer Tanager, Blue Grosbeak, and Grasshopper Sparrow.

Along the Potomac from Carderock to Sycamore Landing   8

Although birding can be rewarding in almost any patch of woods or hedgerow in the month of May, the great good fortune of Washington area birders is that they live along one of the continent's major flyways, the Potomac River. The C&O Canal is an especially satisfactory place to look for the migrants that drop into the treetops along the river and the canal itself. The towpath blesses us all with a stumble-proof route for humans who need to be looking in all directions but down and straight ahead.

Any stretch of the canal from the Capital Beltway (I-495) west is prime habitat year-round for Barred Owls; Belted Kingfishers (except in very cold winters); Hairy, Downy, Red-bellied, and Pileated woodpeckers; White-breasted Nuthatches; and Eastern Bluebirds, joined by Common Flickers, Yellow-bellied Sapsuckers, Brown Creepers, and Rusty Blackbirds in the cooler months. In early spring Wood Ducks move in to nest high in the sycamores, and Eastern Phoebes set up territories on bridges and cliff faces. Pied-billed Grebes and Green-backed Herons start nesting in April and May, mostly west of Great Falls.

The diversity of land birds that arrive to breed along the river is mouth-watering: Yellow-billed Cuckoos; Chimney Swifts; Ruby-throated Hummingbirds; Belted Kingfishers; Great Crested and Acadian flycatchers; Eastern Wood Pewees; Northern Rough-winged and Barn swallows; Wood Thrushes; Blue-gray Gnatcatchers; White-eyed, Yellow-throated, Warbling, and Red-eyed vireos; Black-and-white, Prothonotary, Parula, Yellow, Cerulean, Yellow-throated, and Kentucky warblers; Louisiana Waterthrushes; Common Yellowthroats; American Redstarts; Northern and Orchard orioles; and Scarlet Tanagers. These and a dozen permanent residents range from uncommon and local in their canal-side distribution to abundant and widespread.

It is, however, the concentration of transient species that brings out birders in droves from late April to late May. Some species come through early, some late, and often a species is here and gone in less than a week. Swallows and flycatchers, thrushes and kinglets, vireos and warblers all speed through to the north, offering the local birder only a few days a year in which to observe them in spring plumage and hear them singing. (They return more silently, often in drab and puzzling plumage, from August to October.) Although birding in May is rarely dull, there is no doubt that activity slows steadily to a nadir in midafternoon. You are likely to see and hear twice as many birds between dawn and 9:00 A.M. as during all the rest of the day.

The locations that follow are only points of access to the canal, but they have been birders' landmarks for decades. If you walk west from each, the morning sun will be at your back.

Carderock Picnic Area. Take the first beltway exit north of the Potomac River (marked Exit 41, Carderock and Great Falls), and proceed 1.0 mile west of I-495. There is good birding here on weekdays, but this location is too populated on weekends except perhaps at dawn. Use the left-hand parking lot to bird the edges of the picnic area; use the rightmost lot to reach the canal towpath (20 yards away). Walk west 0.5 mile to the bridge over the canal; just beyond, go left through Marsden Tract Day Use Area. The wooded slopes at the far end are laced with paths down to the river; follow one of the blazed trails clockwise or counterclockwise back to the towpath. There are nesting Yellow-throated and Parula warblers at eye level. You will find an hour or two of fine before-work birding—or more if you continue west to the next area.

Old Angler's Inn. This area is 2.8 miles west of I-495 (same exit) on MacArthur Boulevard. (Go left at the end of the parkway.) Park on the left and walk downhill to the towpath. Look for Cerulean Warblers right there. Walk west on the towpath, making detours down trails toward the river for birds and wildflowers. Look for ducks, hawks, gulls, and sandpipers on or over Widewater, the stretch where the canal expands into a lake. At the second lock west of Widewater cross the canal, and walk back on the park road. Worm-eating

Warblers sometimes nest on south-facing slopes; migrating thrushes and warblers are likely to be seen along the way.

Great Falls Park. Go 5.1 miles west of I-495 (same exit). Bird here weekdays only. (There are mobs on weekends.) Look for ducks in season from the overlook just in front. Orioles like the parking lot edges. Birding can be good east and west along the towpath, and east along the trails through the woods on the north side of the canal. In late winter and early spring cross the canal and go upstream to a cement platform. Walk to the northwest corner of the platform and look across to the island where a pair of Bald Eagles nest. Better views are available at the river edge (and trails go down to it), and you can enjoy the Yellow-throated Vireos and Prothonotary and Parula warblers in the neighborhood.

Pennyfield Lock. Take the River Road exit (Exit 39) west from I-495, 3.3 miles to the traffic light at Great Falls Road in Potomac. Continue 5.1 miles to Pennyfield Lock Road. Turn left and drive 0.9 mile to the parking area by the gate at the end of the road. Cross the canal at the lock just east, and walk west at least 1.5 miles to Blockhouse Point, where the towpath angles sharply right and the river is immediately adjacent.

This stretch can be wonderful in migration—experts can find more than twenty warbler species in a peak morning; novices may find a dozen. The edges of the ponds of the waterfowl sanctuary are noted for shorebirds, bitterns, waterthrushes, Prothonotary Warblers, and Swamp Sparrows. A mile west of the lock, there is a pipeline cut down to the river. Bird the edges of the cut, and, beyond it, look for Scarlet Tanagers, Northern Orioles, and Cerulean Warblers on the far bank of the canal. In migration the cut attracts Black-throated Blue Warblers and Rose-breasted Grosbeaks. Look out over the river from Blockhouse Point in mid-May for hundreds of swifts, nighthawks, swallows, and sometimes a Merlin or two streaming up the flyway. (In winter and early spring it is a good stretch to see Common Mergansers.) If the ground is not too wet, come back along the river's edge, keeping an eye open for Spotted Sandpipers.

Violette's Lock. Turn left off River Road onto Violette's Lock Road, 7.6 miles west of the Potomac traffic light. Check the playing fields on the right for "grasspipers," and bird the rich scrub north of the parking lot by the lock before crossing to the towpath. The big sprawling tree at the parking lot is favored by roosting nighthawks. Take your scope and scan the river just west of here above Seneca Rapids. From autumn to spring, loons, grebes, ducks, and gulls ride the water near the Virginia shore. Migrating Black Terns are often seen in May. Orioles, finches, and sparrows are often especially good here, and warblers and vireos can be abundant. (By the towpath it is 2.5 miles from Pennyfield Lock to Violette's Lock and about 0.7 mile from Violette's Lock to Riley's Lock.)

Seneca. River Road meets a T-junction 0.3 mile west of Violette's Lock Road and continues to the left. After 0.7 mile (8.6 miles west of the Potomac traffic light) go left, just before a flat bridge over Seneca Creek, on Riley's Lock Road through the settlement of Seneca, and park by the lock house in the lot at the end of the road. Cross the aqueduct, below which swallows nest. The next 1.5 miles west can equal Pennyfield in birding delights. Take the spur road that leads off right behind the marshy pond, good for coots, gallinules, and Blue-winged Teal. You can follow an obscure trail, marvelous for warblers, upstream along the north side of the canal for a half-mile or so. (This trail can be reached by car as well. Just west of Seneca Creek turn left on Tschiffley Mill Road and drive to the pond. There is room for only two or three cars.) From the towpath side of the pond you may see Green-backed Herons; Great Egrets; Red-shouldered Hawks; Ruby-throated Hummingbirds; Belted Kingfishers; Eastern Bluebirds; Yellow-throated and Warbling vireos; Prothonotary, Yellow-throated, and Yellow warblers; and Northern and Orchard orioles. Westward along the canal look for waterthrushes, Acadian Flycatchers, and thrushes. After spring migration is over, come back in June and July to enjoy the breeding birds and their young. At the first inkling of autumn it pays to start patrolling the towpath again for the trickle of fall migrants that swells to a flood by mid-September and subsides gradually through October.

Hughes Hollow. About two miles beyond Seneca you will find an entirely different combination of habitats, and thus quite a different assortment of birds from the ones mentioned earlier. From Hughes Hollow west to Sycamore Landing, the species that birders look for include Yellow-crowned Night-Herons (now rarely seen), Least Bitterns, Wild Turkeys, Red-headed Woodpeckers, Willow Flycatchers, Marsh Wrens, Kentucky and Prairie warblers, Yellow-breasted Chats, Blue Grosbeaks, and a wide variety of sparrows.

Hughes Hollow is the birders' name for the part of McKee-Beshers Wildlife Management Area that is accessible from Hunting Quarter Road. This territory begins off River Road 2.8 miles west of the junction with Md. 112. Just before this point on the left is a parking area with good field and edge birding for Prairie Warblers, Yellow-breasted Chats, and Grasshopper and Field sparrows. Turn left on Hunting Quarter Road (an easy road to overshoot). At the first bend, at the bottom of the hill, a gated road runs out to the left. Known to some as the Hunter's Path, this muddy track, together with the adjacent swampy woods, is one of the best sections of Hughes Hollow for thrushes, Kentucky Warblers, and an assortment of migrant warblers. It probably does not pay to walk out into the fields beyond.

It is worthwhile in May to drive along Hunting Quarter Road very slowly with your head out the window for warbler song and your eyes on the road shoulders, looking for Woodcocks, which may with luck be seen at any time of day. Barred Owls are resident; Green-backed Herons, Broad-winged and Red-shouldered hawks, Yellow-billed Cuckoos, Louisiana Waterthrushes, Rusty Blackbirds, and sometimes Purple Finches are all to be expected in their respective seasons. The road can be flooded by a heavy rain and should be avoided if it looks muddy.

The middle section of Hughes Hollow is marked by another chained-off road and a small parking lot, 0.6 mile from the beginning of Hunting Quarter Road. The road goes all the way to the river, mostly through fields where Yellow-breasted Chats and Grasshopper Sparrows breed. Transient ducks are likely to be found in rainpools in the fields. In the wet woods between the fields look for Eastern Bluebirds and Red-headed Woodpeckers; more isolated trees are favored by Eastern Kingbirds, Indigo Buntings, and Blue Grosbeaks. As you continue down Hunting Quarter Road, the swamp on your left becomes a large impoundment favored by Great Blue and Green-backed herons, Great Egrets, Black-crowned Night-Herons, American Coots, Wood Ducks, Blue-winged Teal, and all kinds of swallows. Park at the next chained-off road, 0.2 mile farther along, where there is a large parking lot, and walk out on the dyke straight ahead.

There are three impoundments in all, a huge one on the left and two smaller ones on the right. The one on the left and the first on the right are mostly open water; the second is an overgrown marsh with a rich breeding population of Least Bitterns, Willow Flycatchers (not present every year), Marsh Wrens, White-eyed Vireos, Yellow Warblers, Common Yellowthroats, and Red-winged Blackbirds. The more water-loving species are decreasing as the vegetation matures but it is still an exciting area for both eyes and ears for much of the year. Immature Purple Gallinules once appeared in late summer there. Beyond the impoundments there is a vast area of mixed habitats: fields planted for wildlife, hedgerows, a growing stand of pines and spruces, swampy woods, and brushy tangles. A whole spring day can be spent exploring it for migrants and for summer residents setting up territories; in autumn, sparrows pour in from the north to join the resident species; in winter the conifers attract owls and sapsuckers; and at any time of year it is important to search the sky for hawks and the edges of the fields for Wild Turkeys, present but always elusive. If you go all the way back to the south edge of Hughes Hollow you will discover trails that cross the dry canal bed to the towpath.

Hughes Hollow is always likely to be muddy and is frequently insect-plagued, and the fields and woods are webbed with poison ivy and cat brier and drenched in morning dew—but it is all worth it especially for the novice birder or western visitor. (Note: the impoundment area is more quickly reached by continuing down River Road 1.6 miles beyond Hunting Quarter Road to Hughes Road and turning left. In 100 yards you will be at the west end of Hunting Quarter Road, and another 100 yards east will take you to the parking lot. The field on the left of Hughes Road between River Road and Hunting Quarter Road is especially good for displaying Woodcock from March to May. Across River Road the first quarter-mile of Hughes Road is favored by Blue Grosbeaks.)

Sycamore Landing Road goes left from River Road to the C&O Canal 0.25 mile beyond Hughes Road. A slow drive along it may yield any of the species Hughes Hollow has to offer. Look for Yellow-bellied Sapsuckers and Winter Wrens (from fall to spring) and Brown Creepers. The wet stretch of the road can be wonderful for migrating flocks of warblers that dash across it.

At the end of the road you can cross the canal to the towpath. Right there Warbling Vireos and American Redstarts have their nests, and a hike upriver to your right will lead you past a colony of House Wrens and abundant habitat for woodpeckers and Barred Owls. Sycamore Landing is a prime area for Swainson's and Gray-cheeked thrushes in mid-May. Downriver about a half-mile Cerulean Warblers nest. Continue farther downriver to a campsite, where you will find a path across the canal bed into Hughes Hollow. This stretch of river is as good as Blockhouse Point for watching birds migrating upstream in May. At night, stand here and listen for the Great Horned Owl that calls from a nearby island.

9 Parks in Central Montgomery County

Aside from the area between the Potomac and River Road, Montgomery County has a number of other good sites for birding. Occasionally one of them produces a real rarity and becomes a point of pilgrimage, but in general they simply provide easy access to a variety of attractive habitats.

The locations described in this chapter are all near I-270 between Rockville and the Frederick County line.

Rock Creek Regional Park was developed in the 1970s around two man-made lakes, Needwood and Frank, constructed for flood control of the Rock Creek drainage. Its 1,000 acres contain upland fields, stands of pine, deciduous woods, swamps, and streams. The section

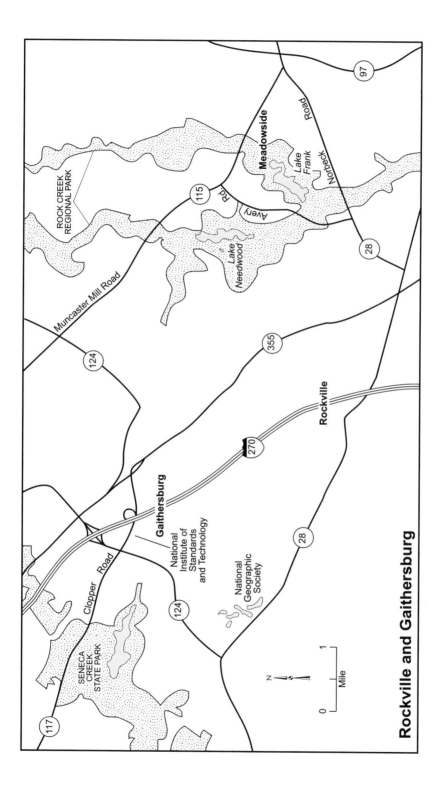

Rockville and Gaithersburg

around Lake Frank and Meadowside Nature Center is managed for fishing, hiking, and nature study.

From I-270 drive east on Md. 28, turning left on this route at 2.5 miles when Md. 586 goes straight ahead. In another 1.4 miles turn left on Avery Road, and continue 0.7 mile to a gravel parking lot on each side of the road, where a bicycle trail crosses the road and a footpath branches off to the left. (The bicycle path runs from Lake Needwood to Washington; the northern terminus is about 1.5 miles to the left.) Birding can be excellent on these trails in migration. If you go on 0.3 mile to the Lake Frank parking lot on the right, the path at the end of the parking lot will take you to the south end of the lake.

Avery Road ends in 1.3 miles at Muncaster Mill Road. Turn right and drive 0.7 mile to Meadowside Lane, the entrance road to the Meadowside Nature Center at the end of the lane, and to the Lathrop E. Smith Environmental Education Center, a school facility. (Alternatively, continue east from the south end of Avery Road on Md. 28 another 2.6 miles to Md. 115, Muncaster Mill Road. Go north [left] on Md. 115 for 1.5 miles and turn left on Meadowside Lane.)

A trail map, a bird list, and staff to answer your questions are available at the nature center, but the center is not open at all on Mondays, before 1:00 P.M. on Sundays, or before 9:00 A.M. the rest of the week. The trails, however, are open from sunrise to sunset. Various feeders, both close to the building and in front of nearby blinds, offer good opportunities to study and photograph resident species and winter finches and sparrows.

Trails of special interest to birders are the Rocky Ridge and Backbone trails behind the center, which provide good eye-level views into the treetops for migrants; the Meadow and Sleepy Hollow trails (the latter with some rough, often wet walking), which have mixed field, hedgerow, and deciduous woodland habitats and are particularly good for thrushes; and the Big Pines and Lakeside trails, which allow you to circumnavigate Lake Frank and the marsh upstream. Owls favor the vicinity of the old graveyard along Walnut Grove Trail, and migrant warblers the clearing around the covered bridge. Bobwhites and Red-shouldered Hawks are among the residents, and Acadian Flycatchers, Brown Thrashers, White-eyed Vireos, Common Yellowthroats, Yellow-breasted Chats, and Indigo Buntings may be conspicuous in summer.

Lake Frank attracts Green-backed Herons and Spotted and Solitary sandpipers in spring and summer, Snowy Egrets in early fall, and any water bird from loons to Oldsquaw in migration. In winter, because of the establishment of hydrilla, a good collection of ducks can be found in Lake Needwood, which can be reached by turning left onto Lake Needwood Drive from Avery Road.

Seneca Creek State Park comprises several discontinuous areas totaling 6,600 acres, mostly along Seneca Creek and its tributaries, with a stretch of managed hunting land between River Road and the C&O Canal. The upper section around Clopper Lake has many visitor facilities, including a good trail system.

To reach this section from I-270 from the south, take the exit for Md. 117, which is Clopper Road, and go west. From the north, take Exit 11B, the one for Md. 124 west. In 0.5 mile turn right on Clopper Road. Drive 1.5 miles to the park entrance, which appears suddenly on the left. Park in the lot for the visitor center (no fee). (Inside the park the fee is substantial at certain times.) If the gate is not open, birders often park along the shoulder of the road a little farther north and bird along the power line or the gas pipeline that cross Clopper Road.

You can pick up a trail map and a bird list in the visitor center; documented reports of additional species or of nesting species will be welcomed by the staff. For your first visit it is a good idea to call in advance and try to join a conducted bird walk. The telephone number is (301) 924-2127. The park opens at 8 A.M. from May to September, at 10 A.M. from October to April.

The recommended birders' route is as follows: take Great Seneca Trail from the northeast corner of the parking lot, with a shortcut along the jeep trail under the power line (White-eyed Vireo, Prairie Warbler, Common Yellowthroat, and Yellow-breasted Chat), as far as the bluff over the creek. Then go back to the Chickadee Picnic Area and work the fields behind the rest rooms (Grasshopper Sparrow and Eastern Meadowlark). Turn right on the main park road and then left to the Kingfisher Overlook, from which you can scan the lake for herons, waterfowl, and shorebirds (Pied-billed Grebe, Common Merganser, and diving ducks in early winter; Green-backed Heron, Wood Duck, and Spotted Sandpiper in summer).

Continue down the main road to the picnic shelters; there may be lots of birds en route. Prairie Warblers and Orchard Orioles are present near the shelters in breeding season, Cedar Waxwings in winter. Then take Mink Hollow Trail, which starts behind the rest rooms; in migration this is an outstanding route for thrushes, vireos, and warblers. After a mile go right on a jeep trail, and cross the meadow diagonally to an old farm road, with hedgerows excellent for field birds. In fall look for raptors overhead and check the trees at the top of the hill for passerine migrants.

Return via the jeep trail to a road over the dam. Turn right on the north side of the lake on the Lake Shore Trail and follow it as far as the Boat Center. Walk through the pine plantation north of it, and turn right on the main road and left through the Cardinal Picnic Area, good for Pine Warblers, Yellow-breasted Chats, and Turkeys. Return to the main road and the visitor center.

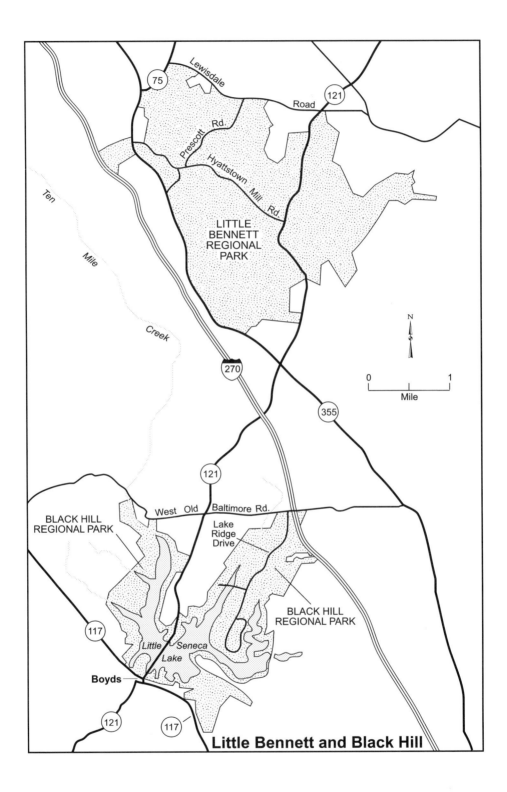

75

Lewisdale

121

Prescott Rd.

Road

Hyattstown Mill Rd.

LITTLE
BENNETT
REGIONAL
PARK

Ten

Mile

Creek

270

N

0   1
Mile

355

121

West Old   Baltimore Rd.

BLACK HILL
REGIONAL PARK

Lake
Ridge
Drive

117

BLACK HILL
REGIONAL PARK

Little   Seneca
Lake

Boyds

121

117

Little Bennett and Black Hill

Two other lakes in the vicinity attract many waterfowl from November to March, but both are open only on weekdays. Continue west on Md. 124 about 0.2 mile beyond Clopper Road, turn left into the entrance to the National Institute of Standards and Technology (known locally by its former name, the National Bureau of Standards), and drive to the lake. Then continue on Md. 124 for 2.1 miles. At Md. 28 turn left and drive 1.0 mile to the entrance to the National Geographic Society. The lake is just ahead. Geese and ducks appear to travel freely between the two lakes, often stopping on the golf course of the Washingtonian Country Club nearby. Look for an occasional Greater White-fronted Goose or a Snow Goose among the Canadas.

You can also reach the National Geographic Society by taking Md. 28 west from I-270 for 4.2 miles.

Little Bennett Regional Park is renowned for the number of its breeding vireos and warblers and its resident Northern Bobwhites and Wild Turkeys. It lies at the top of Montgomery County, some 17 miles up I-270 from its southern terminus; it even spills over a little into Frederick County. From I-270 exit east on Md. 121. At 0.6 mile go left (north) on Md. 355 and drive 0.7 mile to the entrance to the campground and nature center. If you want to pick up a park map and bird list, turn in here if you are visiting on a weekend. Otherwise, continue 2.9 miles, stopping to check the fields to the west for sparrows in winter and Bobolinks in May. Turn right, just beyond a little bridge, on Hyattstown Mill Road and park by the firehouse or continue in your car. Bird down this road as far as you like; there is a parking area at Soper's Branch. Look here for Kentucky and Blue-winged warblers and Scarlet Tanagers. Then retrace your steps, or take Prescott Road west, and you will be back on Md. 355 0.8 mile south of the firehouse. Drive back to Md. 121, and turn left (east). In 0.5 mile turn right into King's Park, a worthwhile stop for a picnic or a walk around the small lake. According to season you may see an American Bittern, a Green-backed Heron, an American Coot, a variety of ducks, Bobwhite, Pine Warblers, Swamp Sparrows, Eastern Meadowlarks, and Northern and Orchard Orioles. (A list as good as this one may take several years to accumulate.)

Continue on Clarksburg Road for 1.3 miles, pull off on the left, and bird along the road. Then go on 0.3 mile to Hyattstown Mill Road, where there is a parking area on the right. Kingsley Trail, which begins there, leads to the Kingsley Schoolhouse, a good stretch for thrushes and breeding Willow Flycatchers; Kentucky, Worm-eating, Blue-winged, Prairie, Pine, and Hooded warblers; Yellow-breasted Chats; and Scarlet Tanager. The meadow adjacent to the parking lot is excellent for watching migrating raptors in fall, as are the hilltop meadows up the trail.

If you turn left on Hyattstown Mill Road you can drive 1.1 miles to a locked gate, and a parking and picnic area.

Black Hill Regional Park lies just west of I-270 off Md. 121. From the exit drive west 1.7 miles on Md. 121 and turn sharp left at the stop sign on West Old Baltimore Road, then a right in 0.9 mile onto the entrance road, Lake Ridge Drive, which makes a long loop through the park. This new park is still being explored by birders and, at this writing, does not yet have a bird list. Little Seneca Lake is already known for its ability to attract migrating loons, grebes (including Red-necked), Double-crested Cormorants, and waterfowl, and there are numerous vantage points over the lake. Barred and Great Horned owls breed in the park, Red-shouldered Hawks and Barn Owls may be resident, and Ospreys, Great Egrets, and herons are regular visitors. In winter look for American Tree Sparrows on the bank above the boat docks. There is also rewarding land-birding along the ten miles of woodland trails. Just south of the entrance station, Black Hill Road takes you west to an arm of the lake and a boat launching area for a different view, with waterfowl on both sides of the road.

When you leave the park, continue west on West Old Baltimore Road 0.7 mile beyond Md. 121 and turn left into Ten-Mile Creek Road. Park on the right before you come to the first "No Parking" sign and explore the trails on both sides of the northwest arm of the lake. Look for Wild Turkey, dabbling ducks and Wood Ducks, and—in winter—for Winter Wrens and American Tree, White-crowned, and Field sparrows. (This area has apparently been birded so far only on Christmas Bird Counts.) Then go back to Md. 121 and turn right (south). In 1.8 miles you will cross the lake on a long bridge and have good views in both directions if you park on the shoulder at the south end of the bridge. Md. 121 ends almost immediately at Md. 117 (Clopper Road). A left turn will take you in 4.7 miles to the entrance to Seneca Creek State Park, on the right.

10 Prince George's and Charles Counties, Maryland

Among the best and most accessible birding areas within the Capital Beltway is Oxon Hill Farm, a National Capital Park at the second Maryland exit east of Wilson Bridge (the first is I-295). Take Exit 3A for Md. 210 south, turn sharp right at the top of the ramp and go right again at the first opportunity. This road will take you back over the

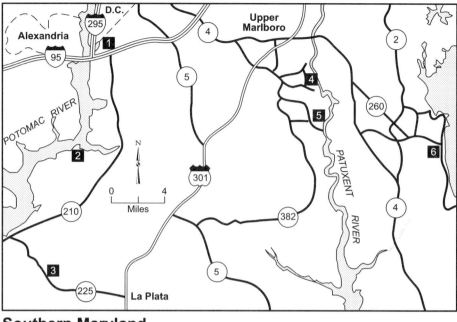

Southern Maryland

1. Oxon Hill Farm
2. Piscataway National Park
3. Myrtle Grove Wildlife Management Area

4. Jug Bay Natural Area
5. Merkle Wildlife Sanctuary
6. Chesapeake Beach

beltway and deposit you in the farm parking lot. If you are coming down I-295, do *not* take the connector to Md. 210—getting back to the farm is practically impossible from there; instead, get onto the beltway heading toward Baltimore.

The first area to cover, especially near dawn during migration, is the hilltop straight ahead, on the north side of the parking lot from the entrance road (leaving the white farm entrance gate far to your left). Warblers and other migrants often drop into the tops of the trees on the slopes below and are relatively easy to study. If the ground is soggy, it is best just to work the woods on your right as far as the power line beyond the old orchard (in plain sight) and come back along the woodland edge on the other side of the apple trees. The orchard is a good place to find woodpeckers and Eastern Bluebirds.

If footing is dry, check the fields to the right of the orchard and beyond it for sparrows. Turn left at the power line, pass the cinder-block building, and follow the trail downhill. The oak-hickory-beech groves may be birdy or silent, but you are at least likely to flush one of the resident pair of Great Horned Owls, and it is a pleasant route to the open fields below. The fields themselves are usually sterile (though falcons, harriers, and meadowlarks are probable in winter), but they are intersected by marshy streams and ponds, especially on your left, with habitat favored by visiting Green-winged and Blue-winged teal and Hooded Mergansers, wintering sparrows, and breeding Marsh Wrens, Yellowthroats, and blackbirds. Look for Willow Flycatchers in the willows. The area often produces Common Snipe in winter or migration.

Ahead, Oxon Run drains into a good-sized lagoon (Oxon Cove), the mouth of which is crossed by I-295. At low tide the creek and lagoon sometimes have migrant shorebirds along the edges and sandbars, and in winter the lagoon harbors coots, grebes, and a wide variety of ducks, especially good numbers of Hooded Mergansers. Gulls are sometimes abundant in the air and on the water, as the Blue Plains Sewage Treatment Plant and the adjacent landfill are just across the way. Flocks should be scanned for rarities, since Glaucous, Iceland, Thayer's, Lesser Black-backed, and Franklin's Gulls have all been reported nearby. This entire area has been productive for Bald Eagles; in winter it is often a good place to see a Peregrine Falcon.

At the lagoon turn left on the paved bicycle path. When it meets a paved road on the left and a gravel road straight ahead, turn right on a grassy path. (The paved road and the grassy path form a stretch where Blue Grosbeaks regularly nest.) At the end, turn left along the edge of a swampy woods, a roosting site for Great Blue Herons, frequented in winter by Rusty Blackbirds, and a likely site for Prothonotary Warblers in summer. You are near the river here, and passerines moving up the Potomac flyway may be dense in the trees and bushes. At the end of the field, come back to the dirt road. Explore the nursery and woods in the southeast corner of the property. The thickets, hedgerows, and fields are excellent for wintering and migrating sparrows. At least nine species including Lincoln's Sparrow have been identified here.

Return eventually to the paved road, which takes you up through the farm exhibits to the parking lot. Alternatively, take the unpaved road that branches off to the left. Just as the old farmhouse comes into sight above you on the hillside, the Woodlot Trail goes into the woods on the left. Look in the Virginia pines for kinglets and nuthatches. Follow the yellow blazes, turn right just past the wooden bridge, and you will end up at the orchard where you started.

The whole area can be covered quite thoroughly in less than a full morning, and travel time from downtown Washington, Alexandria, and much of Prince George's County is negligible.

Piscataway National Park comprises Marshall Hall, the National Colonial Farm, and other land along the Potomac and the south shore of Piscataway Creek on the border of Prince George's and Charles counties. It is best known for its concentrations of waterfowl and Bald Eagles in the colder months but offers good birding all year long. From the beltway go south on Md. 210 for 8 miles and turn right on Farmington Road West. Go one mile to Wharf Road and drive 0.7 mile to Piscataway Creek. After scanning the water and the trees, go back to Farmington Road and continue 0.8 mile to Bryan Point Road. Turn right. In 2.8 miles a road on the right, signed "Piscataway Park, Accokeek Creek" leads to a parking area by the creek. The gate across the entrance road is locked until 9 A.M. A long boardwalk crosses the marsh at the mouth of the creek, where you may find herons, snipe, and breeding Marsh Wrens. Another boardwalk at right angles to the first one leads to Hard Bargain Farm, an environmental study area admitting only groups, only on weekdays. You can explore the fields, marshes, and woods on the way, however. Look for Hermit Thrushes, Winter Wrens, and Swamp Sparrows in winter, Acadian Flycatchers and Prothonotary Warblers in summer, and Barred Owls all year. You can walk along the shore to Mockley Point. Mount Vernon is visible to the west, on the far side of the Potomac. Then continue 0.4 mile to the entrance for the National Colonial Farm; turn in and drive to the parking lot. You may be able to bird here without paying the usual admittance fee. Call ahead, or drop into the visitor center and find out.

Taking advantage of all these points of access to the creek and the river will give you the chance to see as many as twenty-five species of waterfowl in a day, including Tundra and Mute swans and virtually all the bay and dabbling ducks. A Eurasian Wigeon has been found almost annually in recent years and is most often seen from the end of Wharf Road with American Wigeons. In migration, look for Common Loons and Horned Grebes; Red-necked Grebes have been found just across the Potomac and should turn up here as well. Migrating Bonaparte's Gulls often explore the creek. Raptor viewing can be excellent, not only for Bald Eagles and Ospreys, both of which nest along the creek, but for all the other species that commonly live or winter in the area. The farm fields lure in Horned Larks and Eastern Meadowlarks in winter, and American Pipits and Bobolinks in migration.

In spring and early summer Chuck-will's-widows may be found nearby. Continue on Md. 210 past Farmington Road to the light at Livingston Road. Turn right and drive to Old Marshall Hall Road. Turn right and drive 2.4 miles. The woods on the right near the junction with Cactus Hill Road have recently been the best stretch to find the birds. Please heed the notices that the roads in the Moyaone Reserve are closed to the public.

Myrtle Grove Wildlife Management Area is farther afield but still a half-day outing. From the beltway take Md. 210 south for 18 miles to Md. 225, where you turn left and go east for 4.9 miles to the Myrtle Grove entrance on the left. Alternatively, go south on U.S. 301 to LaPlata, turn right on Md. 225, and drive 5.8 miles to Myrtle Grove, on the right.

The entrance road runs for a mile through upland deciduous woods, crisscrossed by trails, to a man-made lake. The lake is popular, when filled, with wintering and visiting grebes and ducks and, when drained, (as it sometimes is, unpredictably) with migrant shorebirds. Bald Eagles are regular visitors. Check the hedgerows along the last hundred yards of the road for White-crowned Sparrows in winter. Blue Grosbeaks breed here.

A dike curves around the left side and far end of the lake; walk the road on top of it until it ends in an area of pines. If it is not too wet underfoot, a visit to the older trees can yield anything from Pine Warblers to Long-eared Owls in season. Then retrace your steps to a road that runs downhill from the dike into a deciduous bottomland. (This road is on the left as you head out along the dike from the parking area.) All along the dike be alert for the eye-level birds in the trees growing below its banks.

At the bottom of the slope go left at the second fork. The loop is perhaps a little over a mile long; to complete it go left when the choice is offered on the far side of the swamp. Along this walk the birds you are most likely to see or hear (depending on season) are Great Blue and Green-backed herons; fleets of Wood Ducks and a few wintering Hooded Mergansers; Red-shouldered Hawks and Ospreys; Barred Owls; all seven of the local woodpecker species, including Red-headed; Winter Wrens; Prothonotary, Parula, Yellow-throated, Kentucky, and Hooded warblers; Louisiana Waterthrushes; Eastern Bluebirds; Hermit Thrushes; and Rusty Blackbirds, as well as an assortment of migrants. Aside from an occasional quiet fisherman relishing the serenity of the swamp as much as you are, you may have the place to yourself. Do not go to Myrtle Grove in hunting season, however, and if you run into riders of horses or bicycles, try again another day.

For those seeking Red-headed Woodpeckers in particular, a good alternative is to drive 13 miles south from the beltway on Md. 210, go left on Md. 227 1.3 miles, and turn right on Md. 224. After this road crosses Md. 344 in 11.6 miles, it narrows and runs through deciduous woods, dipping in several places to swampy areas with dead trees on both sides of the road. This habitat is attractive to the woodpeckers for nesting; two sites where colonies have been found are at 1.5 miles south of Md. 344 and at 6.7 miles farther on, 1.8 miles south of Liverpool Point Road. At the former place eleven Red-headed Wood-peckers and nine Bald Eagles were once found simultaneously! Other breeding species that are likely to be in these locations are

Green-backed Herons, Belted Kingfishers, Acadian Flycatchers, and Parula and Prothonotary warblers.

Jug Bay and Merkle Sanctuary. Not far from Upper Marlboro, the county seat of Prince George's County, a favorite birders' haunt is a stretch along the west bank of the Patuxent River comprising the Jug Bay Natural Area, a 2,000-acre segment of Patuxent River Park, and the adjacent Merkle Wildlife Sanctuary. A recently published checklist shows 100 nesting species in the park alone, plus 154 other species that have been recorded at least once. (Sixteen are considered as accidental.)

A great diversity of habitat, combined with a careful monitoring of public use, make Jug Bay so rewarding that a visit is worth the fee that may have to be paid. This county park requires an annual or daily permit for all individual visitors or an advance reservation for groups, which are admitted free. Annual permits are available only to Maryland residents. The fee is low for Marylanders, lower for Prince George's and Montgomery county residents, and half-price for residents 60 or older. For information on access and fees for out-of-state visitors call the park at (301) 627-6074.

To reach Jug Bay, take U.S. 301 about 1.8 miles south of the intersection with Md. 4, and turn left on Croom Station Road (mile 0.0). At 1.0 mile you will come to a pond on the left, and then a swamp across the road. Park by the bridge at 1.2 miles, well off the road. This is a good spot for Eastern Screech- and Barred owls, Blue-winged Teal and Wood Duck, woodpeckers, and songbirds, including Prothonotary Warblers.

Continue to a T-junction with Croom Road, Md. 382, at 1.7 miles and turn left. Turn left again at 3.2 miles on Croom Airport Road. At dawn or dusk look to the right for the large roost of Black and Turkey vultures in the tall trees behind the house on the corner. Keep right at the T-junction at 4.7 miles.

If you arrive before 8 A.M., the park gate will be closed; continue past the entrance road at 5.3 miles to the end of Croom Airport Road at 6.2 miles, where a field marks the site of the long-defunct airport. The roadsides, the field, and its edges are excellent for such species as Horned Lark, American Pipit, Eastern Bluebird, Prairie Warbler, Yellow-breasted Chat, Blue Grosbeak, and Vesper and Grasshopper sparrows.

The entrance road to the park is worth a stop or two. Try the first dip for Louisiana Waterthrushes and owls, and the birdy second trail on the right for Pileated Woodpeckers.

When you check in at the headquarters, look for recent rarities on the sighting sheet. You should have your scope with you because of the excellent views from the observation tower over Jug Bay and the Patuxent marshes. Twenty-nine species of waterfowl have been recorded here, and the list of long-legged waders, including breeding Least Bittern, is extensive. Look for thrushes along the riverbank and warblers in the trees at the entrance to Patuxent Village.

Below the headquarters, a nature trail and boardwalk will take you to the edge of the superb rice marsh and out to an observation tower. Depending on the season, look for Bald Eagles, Ospreys, Marsh Wrens, Swamp Sparrows, Bobolinks, and Rusty Blackbirds. All the rails, Common Moorhen, and Purple Gallinule are on the park list. While the variety of shorebirds, gulls, and terns is substantial, relatively few species occur annually, much less regularly. Your best bet for seeing most of them is by canoe, available for rent in the park by advance reservation. Local bird clubs and other groups may arrange tours through the marsh in an electrically powered boat with a park naturalist; it may be the best way to see Least Bitterns (in spring and summer) and Soras (in fall) in Maryland. Both the Audubon Naturalist Society and the Montgomery Chapter of the Maryland Ornithological Society regularly schedule such trips.

Inland from the marsh a network of hiking trails crisscrosses the woodland from Black Walnut Creek to Swan Point Creek; take them all if you can. The varied habitats have produced breeding Kentucky, Hooded, Pine, and Yellow-throated warblers; American Redstarts; Yellowthroats; and Scarlet Tanagers, plus migrant warblers and all sorts of sparrows.

One of the most interesting things to do is to take the formidably named Chesapeake Bay Critical Area Driving Tour, which winds five miles through Jug Bay and Merkle land and offers wonderful views over the marsh, the river, and fields and woodland along the way. Informative signboards add to the interest, and there is a stopping point on the long wooden bridge over Mattaponi Creek, a boardwalk, and a forty-foot observation tower where you can get out of your car. (Look for the Summer Tanagers that nest near the bridge.) This tour is open without prior reservations or even a check-in for individuals— and free—on Sundays only, from noon to 3 P.M. Go to the end of Croom Airport Road and turn left. Group tours with naturalist guides are possible at other times by advance arrangement.

When you run out of birds at Jug Bay, it is time to move on to Merkle Wildlife Sanctuary, a good place for an afternoon visit because of its views to the east.

Go back to Croom Road and turn left (mile 0.0). After 1.1 mile, turn left on St. Thomas Church Road, which becomes Fenno Road and then crosses Mattaponi Creek. This creek is regarded by some as the best area for warblers in Prince George's County. Park by the first bridge and walk downstream (away from the left side of the road) on the left side of the creek. Nesting species include Barred Owl; Ruby-throated Hummingbird; Pileated Woodpecker; Kentucky, Hooded, and Parula warblers; American Redstart; and Louisiana Waterthrush, supplemented by other species in migration.

The Mattaponi divides Jug Bay from Merkle. At 2.8 miles from Croom Road turn left at the entrance to the sanctuary and drive 0.9 mile to the visitor center. The area is open from 7 A.M. to 5 P.M. The visitor center, open from 10 A.M. to 4 P.M., has a variety of programs for families and children, instructive exhibits, well-placed decks, and a huge indoor observation window overlooking the geese flocks. Pick up a map of the sanctuary there. It shows the three hiking trails into different habitats. Bird walks are regularly on the calendar, and special ones can be arranged for groups. Visitors are encouraged to call the visitor center at (301) 888-1410 to confirm dates and hours of operation.

The area is managed for Canada Geese, and thousands of them are often present from November to February. Other waterfowl join them, of course, and the open land is much hunted over by raptors. It is a good place to see the Ospreys (absent in winter) and Bald Eagles that live nearby, as well as the three breeding buteos: Red-tailed, Red-shouldered, and Broad-winged hawks.

The Piedmont

There is hardly a better area to bird west of Chesapeake Bay than southern Frederick County in Maryland. The wooded slopes around Sugarloaf Mountain, the fish ponds of Lilypons, the upland fields and pastures off New Design Road, and the back roads and trails along the Monocacy and Potomac rivers provide a remarkable diversity of habitat. On a fine day in May it is not hard to find one hundred species in the course of twelve hours or so, if you reach the foot of Sugarloaf Mountain early enough to hear the Great Horned Owls, Eastern Screech-Owls, and Whip-poor-wills.

Leave I-270 at the Clarksburg exit (mile 0.0), drive 0.5 mile northeast to Md. 355, and turn left. At 1.5 miles go left again on Comus Road and follow it to the foot of Sugarloaf Mountain at 7.0 miles, listening for night birds the last two miles or so. Park at the circle and bird on foot the edges of the property, officially named The Stronghold. There is a large grove of conifers just inside the entrance and along the last leg of Comus Road just beyond the circle.

All the migrant thrushes and many species of warblers show up here and along Sugarloaf Mountain Road, which runs northeast from the circle. This stretch has been known to produce migrants in spectacular numbers and variety on gray days when the ceiling is so low that the mountain itself is invisible.

As soon as the gates open at 8 A.M. (before the place is overrun with tourists), drive up the mountain and sample any of the trails, looking especially for Summer Tanagers and Worm-eating Warblers on the lower slopes. Hawk watching from the summit, spring or fall, can be quite good, but the climb is steep and anything but solitary.

Comus Road runs along the left side of The Stronghold past a pond to the west and ends at Mount Ephraim Road (0.4 mile), a narrow, unpaved road that is excellent for woodland birding. If you follow it north (right) to its end on Park Mills Road, 4.2 miles from the circle, you will be just 1.0 mile north of Lily Pons Road.

A longer but equally interesting route from Sugarloaf Mountain (mile 0.0) to Lilypons (the name of the place is one word, the name of the road is two) is via Sugarloaf Mountain Road to Thurston Road at 2.5 miles and Thurston Road north to Peters Road (marked by an ill-placed sign) at 4.9 miles. Make a hairpin turn left on Peters Road, which is pretty and mostly unpaved, following a stream through a narrow valley, with a good chance for ducks, kingfishers, orioles, and sparrows. At 7.3 miles you will have climbed out of the valley to intercept Park Mills Road, where you turn left toward Lily Pons Road.

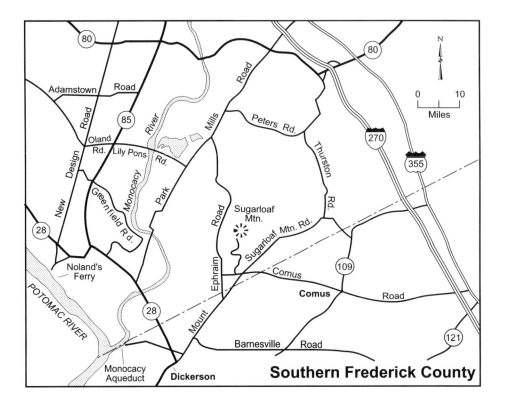

Southern Frederick County

If you take this route or Mount Ephraim Road slowly, you should have no trouble at all, but both are narrow gravel roads for most of their length. If you are more comfortable on broad, paved surfaces, leave Sugarloaf via the southwest on a road that soon becomes Mount Ephraim Road and reaches Md. 28 on the north side of Dickerson at 2.8 miles. (In May, look for Bobolinks in the alfalfa fields.) Go north on Md. 28 for 3.7 miles to the junction with Md. 85 and continue north on Md. 85 for 2.6 miles to Lily Pons Road. Turn right, drive 1.3 miles, and, after crossing the Monocacy River, park near the entrance. (A short detour down Criss Ford Road 0.45 mile from Md. 85 takes you along a little creek often frequented by a pair of Louisiana Waterthrushes.)

Lilypons Water Gardens—named after the opera singer, who came to its opening—is an extensive network of diked ponds where water lilies and goldfish are raised. The water levels and degree of maintenance of the ponds fluctuate constantly. Some are full, attracting waterfowl in migration; some are drained, pulling in American Pipits and a good assortment of shorebirds; and others provide enough reedy cover to entice a regular breeding population of Least Bit-terns, Marsh Wrens, and King Rails. Green-backed Herons, Red-winged Blackbirds, swallows of all kinds, Belted Kingfishers, Ospreys, Black and Turkey vultures, and Red-tailed and Red-shouldered hawks are plentiful in season. From time to time a Loggerhead Shrike or two takes up residence. Ibis, egrets, and herons often show up in late summer, wandering inland from coastal breeding colonies.

The direct route from Washington to Lilypons is to take the Urbana-Buckeystown exit (mile 0.0) from I-270, heading west on Md. 80 for 1.7 miles to Park Mills Road. Turn left and follow Park Mills Road 5.3 miles (past Peters Road at 3.7 miles and Mount Ephraim Road at 4.3 miles) to Lily Pons Road.

Lily Pons Road has wide, paved shoulders for parking. The en-trance road to the sales office is 0.6 mile from Park Mills Road, but birders are asked not to use the small parking lot for customers, and the limited hours that the gate is open are restricting anyway. (The area is open to birders on foot at all hours.) As Lily Pons Road approaches the bridge over the Monocacy, it gives you an excellent view over all the ponds on the south side of the property. You can check for water levels and the location of shorebirds or ducks before you set out on foot.

Don't go to Lilypons when the air is hot, heavy, and still; it is an exceptionally miserable heat trap. Otherwise it is almost always interesting, at least from March to November.

West of Lilypons, Md. 85, which runs north to Frederick, is a heavily traveled road with no shoulders to park on. Parallel to it and 0.7 mile to the west lies New Design Road, a wide, quiet farm road, renowned for its nesting Grasshopper and Vesper sparrows and Horned Larks, its migrating American Pipits, and its wintering Rough-legged Hawks and Short-eared Owls. Lapland Longspurs, and, more rarely, Snow Buntings are sometimes in the wintering flocks of Horned Larks. The birdiest part of New Design Road is the 2.2-mile stretch from Oland Road south to Md. 28.

Oland Road, the westward extension of Lily Pons Road to New Design Road, is bordered on the south by a pasture where the sandpipers, pipits, and sparrows are often most easily visible. Look for Upland Sandpipers there in late July and early August.

South of Md. 28, New Design Road leads in 0.75 mile to Noland's Ferry, a picnic spot between the C&O Canal (here dry to damp) and the Potomac. Though best in the morning, it can be good all day in migration for birds on the river flyway: hawks, ducks, gulls, swifts, swallows, and nighthawks, as well as woodland species along the towpath, especially flycatchers, tanagers, and orioles, and the resi-dent Barred Owls and woodpeckers.

You can also reach the canal by taking Mouth of Monocacy Road off Md. 28 0.3 mile north of Mount Ephraim Road and 3.4 miles south of the end of Md. 85. Keep left at the fork to reach Monocacy Aqueduct, a handsome bridge of white granite. These two points of access to the canal are 2.4 miles apart.

If you have time you might explore the 3.3-mile loop made by Greenfield Road, heading east from Md. 85 a mile south from Lily Pons Road. It goes south along the Monocacy, past long-abandoned fish ponds, through woods and fields, returning to Md. 28 0.6 mile below its junction with Md. 85.

If you are cruising the roads of this area in winter in a quest for Snow Buntings and Lapland Longspurs, a road worth checking is the one up to Claggett Diocesan Center, on the east side of Md. 85 3.0 miles north of Lily Pons Road and 0.1 mile south of the junction of Md. 85 and Md. 80, 5.1 miles west of I-270. The best conditions are to be found a day or two after a snowstorm, when the roads have been plowed and the fields are still blanketed. The birds will be on the road or the shoulders. You may bird the grounds of the center after checking in at the office, but the entrance road is likely to end your quest.

12 Loudoun County, Virginia

Among the more accessible and rewarding areas in the northern Virginia piedmont are several locations in Loudoun County. The sites described in this chapter can be combined into a long morning or a short day of pleasant open-country birding at almost any time of year. The county is rapidly being paved over, however, and all the locations described here that are not public land may be transformed into subdivisions in the next few years.

Note: If you buy a map of Loudoun County, the road names on it may differ from the ones at the intersections. When in doubt, go by the road *numbers*.

Starting from the intersection of the Capital Beltway (I-495) and Va. 193, drive west on Va. 193 for 9.3 miles until it merges with Va. 7 at Dranesville. Continue west on Va. 7 for 2.9 miles and turn right on Road 637 at the sign for **Algonkian Regional Park.**

This small park on the banks of the Potomac is clearly managed primarily for golfers, swimmers, picnickers, and boaters, and is likely to be crowded with all of them in pleasant weather, especially on summer weekends. From January to March you may have it to yourself all day, but during the rest of the year it is advisable to arrive close to dawn and plan to spend no more than a couple of hours there.

Some of the best birding is along the river; turn left at the T-junction and drive to the picnic area left of the boat-launching area. The Potomac itself should be scanned for Ospreys from March to October, and for loons, grebes, geese, diving ducks, and mergansers in the colder months; it serves as a highway for migrating nighthawks, swifts, and swallows in spring and fall. A walk from the picnic area to the chalets upriver and over to the swimming pool can be especially good for woodpeckers and warblers. The entrance road itself may offer good birding during migration on a weekday morning when there is not much traffic.

A right turn at the T-junction leads to the parking lot for the golf course. (The pro shop behind the conference center has public rest rooms.) The strip of trees along the left side of the road is good for Northern and Orchard orioles, Scarlet Tanagers, treetop warblers, and perhaps Yellow-throated and Warbling vireos. The marshy area on the far side of that strip of trees has turned up Common Snipe in season, and is, incidentally, of particular interest to botanists for the freshwater meadow plants and wildflowers that grow there.

Notice the "Do Not Enter" sign on the way to the pro shop. This blocks the former entrance road, by which you can reach on foot the far side of the golf course without bothering the golfers. This area has nesting Prairie Warblers and Yellow-breasted Chats and migrating Blue-winged and Golden-winged warblers. It is best birded from the sewer line along the park border, which is good later in the morning after the park fills with people.

Red-shouldered Hawks and Barred Owls are resident around the park, and so are Eastern Bluebirds, thanks to plenty of bluebird boxes (which are also popular with Tree Swallows), but the bluebirds are most likely to be seen close to dawn or dusk.

When you run out of birds or peace at Algonkian, go back to Va. 7. Head west (right) on Va. 7 for 10 miles and take the Leesburg bypass on U.S. 15. Take your mileage at the intersection.

In 0.3 mile Road 773, a three-sided 4.7-mile loop road that returns to the bypass 0.4 mile farther on, takes you quickly away from the traffic's roar to a quiet, untraveled, unpaved woodland road along the second side, as yet unexploited by developers or even birders. The third side, Edward's Ferry Road, passes Red Rocks Wilderness Overlook Regional Park, a nature study area on the banks of the Potomac with a mile and a half of trails. Try this area in migration— you may have it all to yourself.

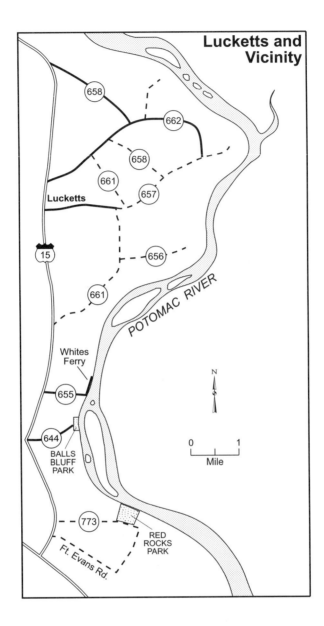

The next road to the right leads to the smallest of all our national cemeteries, Ball's Bluff, worth a short visit both in its own right and for its little network of occasionally birdy riverside trails.

At 3.6 miles (from Va. 7 without a detour on Road 773) you will pass the road to White's Ferry (which you may take to western Montgomery County, Maryland); then you will cross Limestone Branch and come (at 5.0 miles) to Road 661 on the right. Turn here and slow to a crawl. You are entering a section of farmland, behind the village of **Lucketts,** laced by almost untraveled gravel roads where you can stop virtually at will if you just pull over far enough to allow another car to pass. Please do not trespass anywhere.

The area is especially good in the appropriate season for Vesper and Grasshopper sparrows, soaring hawks (including Northern Harriers) and vultures, American Pipits and Horned Larks, Bobolinks in May, and sometimes a Loggerhead Shrike or two. The lush pastures with scattered rocky outcroppings should be scrutinized for Upland Sandpipers. (You may just see a head above the grass.) Farm ponds, hedgerows, and a few patches of deciduous woods all are worth checking, and the occasional little streams can produce anything from a Louisiana Waterthrush to a Yellow-crowned Night-Heron.

A small bridge crosses a creek 0.4 mile from the highway. Park just beyond it on the left and scan the woods on the left bank for the resident Great Horned Owls and the dead snags for Red-headed Woodpeckers. The next farm on the left is due to become a regional park sometime soon, perhaps by the time you try this route. A right turn onto Road 656, 2.2 miles from U.S. 15, will lead to a dead end at a farm after going through bottomland pastures and fields. In winter look in the low-lying wet meadows (on this and all the other roads in the area) for Common Snipe. Try to keep out of the way of farm trucks, and turn around at the cluster of mailboxes.

Road 661 ends at a T-junction with Road 657, where a left turn will take you back to U.S. 15 and a right turn will eventually (in 3.0 miles) lead to another dead end. (Turn around when you come to a sign for a nursery.) Retrace your route to the first right turn onto Road 662. In 2.2 miles go right on Noland's Ferry Road. At its end an overgrown trail leads down to the river. Continuing on Road 662 another 0.3 mile, turn right again onto Road 658, which ends at U.S. 15 after 2.0 more miles. The last half-mile or so of Road 658, on both sides of the road, has been the best area to see Upland Sandpipers. An alternate area to look for them is the field that lies along the north side of Road 662 west of Road 658 and east of Road 661.

Similar habitat and back roads lie on the other side of U.S. 15 and on the south side of Va. 7 between Sterling and Leesburg, and you should explore them for yourself, preferably armed with a county map. Try the W&OD Trail, described in Chapter 7.

Finally, **Beaverdam Reservoir,** known to many local fishermen but as yet to very few birders, is a clear, quiet lake northwest of Dulles Airport well worth a visit from late October to April, when loons, grebes, and waterfowl drop in during migration and may stay until the water freezes over. As the source of the water supply for the City of Fairfax, it is forbidden to swimmers, hunters, and fishermen with gasoline-powered boats. Parking is very limited. For a morning visit (that is, to see it from the east) take Road 659 south from Va. 7 for 5.6

miles south to Road 625, Mount Hope Road, and turn right 0.1 mile to the lake. Another vantage point lies 0.4 mile south on Road 659 at the end of Road 646, Mount Middleton Road (0.4 mile from the highway).

Crossing from Road 659 to Road 621 via Road 772 can be rewarding. Check the fields near the intersection with Road 659 for Horned Larks and American Pipits. Be alert for Wild Turkey anywhere along this road, especially near wooded areas. The woodlots may yield a good variety of sparrows, in winter.

You can reach the reservoir from the south by taking U.S. 50 west of Va. 28 for 4.9 miles and turning right on Road 606. In 0.9 mile fork left on Road 621, Evergreen Mills Road, and turn right on Road 659 1.6 miles farther on. (You will be 3.6 miles from Road 646.) To see the lake from the west continue north on Road 621 for 4.9 miles and turn right on Road 861, Reservoir Road. This peaceful country road runs past farms to reach the lake in two miles. Road 621 ends at U.S. 15 at the west end of the bypass around Leesburg, 6.2 miles north of Reservoir Road.

13 Lake Anna

In a triangle the corners of which are marked by Charlottesville, Richmond, and Fredericksburg, Virginia, lies long, man-made, and many-armed Lake Anna, warmed by the discharge of the cooling canal of a nuclear power plant and enormously attractive to waterfowl. It is an especially rewarding area to explore in late fall and early spring.

Essentially, the route that follows begins at the west end of the lake and runs east along the south side, with several detours north to fingers and bays along the lake shore.

To reach the starting point take I-95 south to Exit 45-B near Fredericksburg. Go west 12.4 miles on Va. 3; you will pass the Chancellorsville Civil War Battlefield. Turn left at the traffic light onto Va. 20, and follow it 13.6 miles to U.S. 522. Take your mileage at this intersection. Head south on U.S. 522 and in 8.8 miles make your first stop where the highway crosses unmarked Pamunkey Creek. This initial drive from Fredericksburg is an attractive route past rolling piedmont farms; you should be on the alert for raptors all the way. A particularly productive area for hawks, including the occasional Rough-legged Hawk and Northern Goshawk, can be reached by turning right on County Route 651, then left on Route 649 and back to U.S. 522.

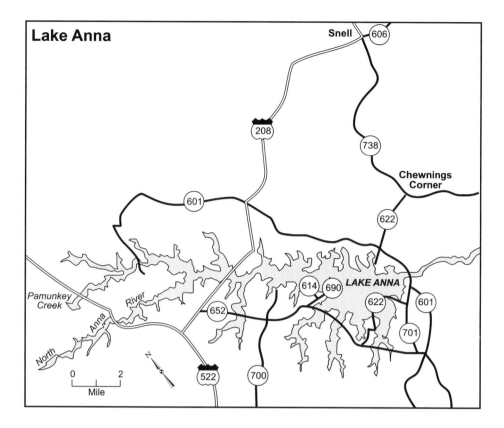

Pamunkey Creek can be good for gulls, ducks, and shorebirds in season, especially in the cove west of the highway. You can park by the dumpster on your right or at the Virginia Game and Inland Fisheries boat ramp on your left. In fall and winter Bald Eagles are often seen on the snags in the cove.

You cross the North Anna River next, then at 14.7 miles turn left on Va. 208, which crosses the lake itself. You can scan the water briefly from your car on the shoulder or park (with permission) at a campground on the northwest side of the highway. Look for loons, diving ducks, and mergansers, as well as gulls and terns. Another good vantage point can be reached by turning left at the far end of the bridge (18.1 miles) into the motel and marina area.

Make a U-turn there, retrace your path, and turn left on Road 652 (21.2 miles). The next left, at 25.2 miles, is Road 700. This will take you to the nuclear plant and North Anna Visitor Center (open Monday to Friday), which you may find instructive. On the coldest days of winter, the stops to the west of this point often are mostly frozen over, but all stops to the east are open. Clearly, the warming effect of waters flowing out of the plant through the cooling canal is substantial.

Turn left from Road 652 again, 1.9 miles east of Road 700, onto Road 614 and explore all its side roads except for those marked by "No Trespassing" signs. Road 614 itself turns left and ends at a birdy cove called Harris's Point; you may park at the "End of State Maintenance" sign (28.6 miles). American Tree Sparrows are often present in winter just past the parking area. Return to the junction of Road 614 and Road 690 (29.2 miles). Turn left, and take 690 to its intersection with Road 1201 at a sign for Both Waters Estates. Lake Anna Road is to the right. A left turn on Road 1201 leads to Duerson Point, with views in three directions and consistently more birds than any other spot on the lake. In this residential area you may park on the shoulders to scan the water or request permission from homeowners to walk to the shore.

Wintering birds include numerous Pied-billed Grebes, Coots, American Wigeon, Greater and Lesser scaup, Bufflehead, Common Goldeneye, and Hooded Mergansers. The records show that plenty of puddle ducks, Canvasbacks, Redheads, and Ring-necked Ducks, as well as Common Loons, Horned Grebes, occasional Red-necked Grebes, Double-crested Cormorants, gulls, and terns have turned up here as well; it is one of the best places on the lake to look for Bonaparte's Gulls. A variety of finches and sparrows, including White-crowned Sparrows, can be flushed out of the brush or seen at the feeders.

Return to Road 652 and reset your odometer to 0.0 miles. Turn left on 652, make an almost immediate right onto Road 1212, and continue to the end of the road for more good views of open water. Come back to Road 652 (0.6 mile) and turn right.

At 3.6 miles turn left on Road 622. Look for Red-shouldered Hawks at the corner and check the edges of the road for land birds. Where the highway bears right at 4.5 miles, turn left onto Road 1271 into a development called Jerdone Island. You can survey a cove from the boat launching ramp at the west end of the road *if you have home-owner permission.*

Continuing on Road 622 for another 0.5 mile, stop at the pond for herons and egrets and check the snags in honor of the Olive-sided Flycatcher that was there one spring. You will then cross a long causeway with the main dam of the lake on the left. You can bird from your car or park at either end. In addition to all kinds of ducks, including occasional scoters, look for loons, Horned and Pied-billed grebes, Double-crested Cormorants, and Bonaparte's Gulls, and hope for Rough-legged Hawks and Snow Buntings, the latter often present along the road shoulders or on the rocks here. In the warmer months, especially in autumn, egrets, terns (including Black Tern), and shorebirds are all possibilities.

The most direct route back to Washington is as follows: take Road 622 at the east end of the causeway (mile 0.0) to Road 701. Stop here in spring and summer for Blue Grosbeaks and in winter for a possible Northern Goshawk or Rough-legged Hawk. Then go left on Road 701 at 0.7 mile and left again on Road 601 at 1.6 miles. Check the pond on the right for puddle ducks and herons and egrets in summer and fall. Go right on Road 622 (Fairview Road) at 4.2 miles and left on Road 738 (Partlow Road) at 7.0 miles to Snell at 16.1 miles. In Snell turn right on Road 606, which will take you to I-95 in 5.0 more miles.

Alternatively you can continue straight on Road 601 and then turn left into Blounts Harbor for a good view of a part of the lake where loons, grebes, and mergansers are often present. It is the only reliable site on the lake for American Coot. Back on Road 601, continue for 1.5 miles to Road 614, turn left, and proceed to Dukes Creek Marina. Park here and scan the lake. Bald Eagles are often present in winter. Retrace your steps to Road 601, continue for 4.5 miles, and turn right on Va. 208. After 17 miles turn right on U.S. 1, which will take you to I-95.

Most of the field work around the lake has been done from fall to spring, with the greatest numbers of birds showing up from November to March. Perhaps the best times to visit are in November and March, when waterfowl are on the move, or during some cold winter period when the Potomac has frozen over and ducks are looking for open water. Because the lake is quite new (it was first filled in 1972), the habitat is still changing, with marshes and reed beds forming around the edges and likely flats for shorebirds developing here and there, especially at the west end.

Lake Anna will repay exploration at any time of year, including the day after a hurricane! Included among its most exotic visitors are Sooty Shearwaters, Brown Pelicans, and Sooty and Bridled terns.

Ridges, Valleys, and the Appalachian Plateau

In early May the variety of birds is so rich and the weather typically so perfect that most Washington birders have no need to go far from home to find all the woodland species anyone could ask for. Later in the month the first sticky day may bring with it a distraction of midges along the canal, a reminder of the drawbacks of a Washington summer, and a stimulus to head for the nearest hill.

From mid-May to the end of June the Shenandoah National Park, less than 70 miles from the Capital Beltway, offers the opportunity to study northern migrants, high-country summer visitors, and permanent residents hard to find closer to home. Not only that, but the park has a recently compiled bird list that is available at the Big Meadows Visitors Center, where a sighting log is kept up to date by both staff and transient observers. Since the published bird list is based on an all-too-thin set of records, a copy of your day's list on any visit would be enthusiastically welcomed, especially if you included numbers and locations for all but the commonest species.

Because of the park's immense popularity you will want to be within its boundaries and birding as close to dawn as you can. Luckily the radio-toting multitudes are not on the whole early risers, but day hikers may be.

By the time you get to Warrenton you should decide how you will spend the morning, especially if you are making the trip on a weekend. For migrants, including raptors, take U.S. 211 toward Thornton Gap. For nesting warblers and vireos, stay on U.S. 29 to Orange and there turn southeast on Va. 230, which joins U.S. 33 at Stanardsville. U.S. 33 takes you west to the park entrance at Swift Run Gap.

If you choose the former route in May and start birding as soon as you leave Sperryville, pulling off wherever you can find a spot, you should be able to see and hear over forty species before you reach Skyline Drive.

Turn south on the drive and stop at every overlook and picnic area that catches your eye. The most widespread breeding species near the road will be Chestnut-sided Warblers, Rose-breasted Grosbeaks, and Dark-eyed Juncos. Ravens are likely at any stop (all year), Solitary Vireos probable anywhere above 2,500 feet.

The figures after place names in this chapter are mileages based on the mileposts along Skyline Drive.

At both ends of Mary's Rock Tunnel (32.4), look and listen for Winter Wrens and Eastern Phoebes; at Hemlock Springs Overlook, walk along the shoulder, checking the hemlocks for Black-throated

◀ Golden-winged Warblers

Blue and Black-throated Green warblers, as well as winter finches—Pine Siskins may linger late when they are around at all.

Take note of the parking area for White Oak Canyon Trail and a third of a mile farther along turn left into the unmarked parking area for Limberlost Trail (42.9), a mile-long loop that yields Solitary Vireos, Blackburnian Warblers, and Veeries as nesters, plus unpredictable transients.

If you take a sharp left just beyond the bulletin board you will link up with a bridle path that runs through an old orchard and second-growth woodland. Here you might find Ruffed Grouse if you are the first on the trail for the day.

Grouse are the park's star attraction for lowland birders, but the difficulty of finding them is enough to spoil your day unless you set your expectations very low. The only strategy to follow is to enjoy any bird you may find in the habitat that grouse like—scrubby cutover land with a mixture of open ground and small trees with low branches.

The next stop is the parking area for the upper Hawksbill Mountain Trail (46.7). The mile-long trail to the summit is good for migrants, and the brush across the drive holds nesting Woodcock and grouse in the low hemlocks. The trail up the north side of Hawksbill is shorter and steeper, but good for Canada Warblers.

By now, if you have been birding systematically, it is likely to be lunchtime just when you find yourself at Big Meadows Wayside (51.3 miles), where you can buy lunch if you like and stop in at the visitor center for bird news.

Save birding at Big Meadows for the afternoon; the open-country birds there are likely to be visible at any hour and the area is enough off the beaten path that you may have it to yourself.

Head across to the east side of Skyline Drive and work around the edge of the meadow, pushing back into the brush wherever it opens up enough. You are as likely to see not only grouse but also Wild Turkeys here as anywhere, but the easy species will be Field and Song sparrows, Eastern Bluebirds, and Common Yellowthroats. Irregular nesters have included Prairie Warblers, Least and Willow flycatchers, Blue Grosbeaks, and Vesper Sparrows. Migrating Common Snipe are out in the boggy center of the meadow.

At this point you have reached the end of both this route and the alternate one coming up from the south, described next.

If you approach the park from Swift Run Gap, at the tollbooth you will be 2.8 miles south of the South River Picnic Area (62.9 miles), the trail head for the best walk for breeding birds in the park. Since it is also a popular trail for hikers, get there as soon after sunrise as you can.

With plenty of birds, this 3.5-mile loop can take you most of the morning. Your route, the falls trail, will take you downhill 1.0 mile to an overlook across from the top of South River Falls, then up to the left on a trail that connects with the South River Fire Road, which deposits you on Skyline Drive just north of the picnic grounds. (You can continue down to the foot of the falls for the view, but it is steep down and steep up; don't say you weren't warned!)

The particular charm of this walk is a result of the steep slope across which the trail runs, allowing good views into the forest canopy. The number of breeding species is remarkable: Louisiana Waterthrush; Ovenbird; American Redstart; Black-and-white, Blackburnian, Hooded, Worm-eating, Black-throated Blue, Black-throated Green, Parula, and Cerulean warblers (the last mostly on the fire road); Red-eyed, Yellow-throated, and Solitary vireos; Wood-Thrushes; Veeries; and Scarlet Tanagers.

Up the drive, milepost 61 marks one of the best areas for Ruffed Grouse; see what else you can find there. The next good stop is at Booten's Gap (55.3). The parking area is beside the Appalachian Trail, which meets Laurel Prong Trail 0.5 mile to the north. Turning right on Laurel Prong, you should find several pairs of nesting Canada Warblers in the first mile.

The entire stretch from Milam Gap (53.0) to Tanner's Ridge Overlook (51.7) is excellent birding, easily accessible because of the Appalachian Trail, which runs west of and parallel to Skyline Drive here. Grouse are always possible, and this may be the best area in the park for migrants, even better in fall than spring. Big Meadows lies just ahead.

Although the focus so far has been on spring and early summer, the park can be just as rewarding in the fall. Land-bird migrants stick to the ridges and peaks, and hawk watching can be excellent. The rule is to look on the side of the ridge from which the wind is coming. (Don't look for hawks on calm days!)

The best all-wind hawk-watching spot is Hawksbill, which offers a 360° view. For westerly winds, Mary's Rock, up a 1.8-mile trail above Panorama at Thornton Gap, and Stony Man, a mile walk up Stony Man Nature Trail close to Skyland Lodge, are prime spots, and The Point Overlook (55.8) can provide superb viewing from your car.

Big Meadows attracts harriers and falcons irregularly and in small numbers. On a cold day you can scan it comfortably from inside the visitor center, but you should not expect a sure reward. Later in the season Snow Buntings may drop in, and winter finches may be found in the spruces north of the center.

Finally, a word on chickadees. Most of those you see in the park will be Carolinas and virtually all the rest will have features of both Black-caps and Carolinas. Though pure Black-caps have been recorded in the park, their presence these days is very much open to question. Reputable birders check every field mark with great care and still leave many chickadees unidentified.

15 Highland County, Virginia

Almost anyone with wanderlust is pulled toward the mountains in October, and you may want to explore Highland County in western Virginia for the first time while the foliage is at its most beautiful and the migrants are slipping south along the ridges.

The resident Ruffed Grouse, Common Ravens, Black-capped Chickadees, and Red Crossbills may be seen at any time, though, and midwinter and early June are even more inviting. From December to February or March, Golden Eagles are always present, and the nesting species of early summer include Winter Wrens, Hermit Thrushes, Golden-winged and Mourning warblers, Bobolinks, and Rose-breasted Grosbeaks.

The most direct route from the Capital Beltway, I-495, is via I-66 and I-81, a swift, dull 2.5 hours to Staunton. From I-81 drive west on U.S. 250 about 22 miles, and, in spring, stop along the road for the next mile or two to listen and look for Golden-winged Warblers.

At 24.9 miles Ramsey's Draft Picnic Ground is a good spot to look for Black-capped Chickadees, breeding Winter Wrens, and Black-throated Green and Parula warblers, either next to the road or along the hiker's trail.

The highway to the west becomes a very steep, winding road with many switchbacks. Exactly 1.9 miles west of the picnic ground turn-off, as the road climbs Shenandoah Mountain it curves sharply left. Park at the curve on the broad gravel shoulder on the right. An hour or so after sunrise at any time of year, Red Crossbills come to collect grit at this curve after their first meal of the day. They are quite tame, and you can usually get out of your car and watch them scratching around on the opposite shoulder. Some may be in the pines on either side of the road. They cannot be relied upon at other times of the day.

Continue 0.1 mile to the crest of the mountain and the county line (Highland County to the west, Augusta County to the east). Park in the lot for the Confederate Breastworks, if you like, and walk up the trail, which is steep for the first half-mile and then level. The first hikers of the day may find turkeys and grouse, while Dark-eyed Juncos, Pine Warblers, and Red Crossbills are among the nesting species.

Red Crossbills may, in fact, be found all along this ridge, north from U.S. 250 to U.S. 33 in Rockingham County. If you did not see them on the eastern slope, look for them on the western slope, just before the sign "Highland FFA Welcomes You."

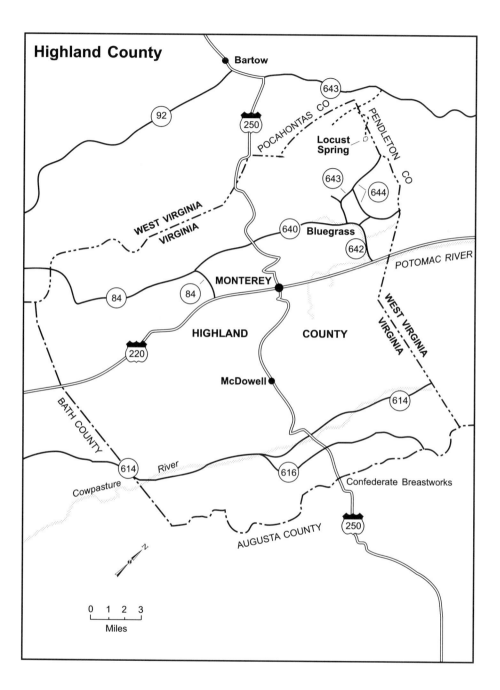

Highland County

Bartow

643

92

250

POCAHONTAS CO

PENDLETON CO

Locust
Spring

643

644

WEST VIRGINIA

VIRGINIA

640   Bluegrass

642

POTOMAC RIVER

84   84   MONTEREY

WEST VIRGINIA

VIRGINIA

HIGHLAND   COUNTY

220

McDowell

614

BATH COUNTY

614   River

616   Confederate Breastworks

Cowpasture

250

AUGUSTA COUNTY

N

0 1 2 3
Miles

From the crest of the mountain, drive 2.9 miles and turn left on Road 616. A delightful loop of 7.1 miles returns to U.S. 250 via Road 614, turning right at the T-junction immediately after crossing the Cowpasture River on a small right-angle bridge, rejoining the highway less than 2 miles west of Road 616. The lack of traffic makes it easy to stop to bird wherever you like, but the most unusual species to a piedmont birder is likely to be a Black-capped Chickadee.

The next ridge you climb is Bullpasture Mountain. Stop on the eastern slope wherever you can park off the road to listen and look for the nesting species: Yellow-throated and Solitary vireos, and Kentucky, Cerulean, Golden-winged, Worm-eating, and Parula warblers.

In the valley on the far side is the village of McDowell, a good spot for Eastern Bluebirds, which nest in the fenceposts; from there to the top of Jack Mountain, the next ridge, Golden-winged Warblers and Cedar Waxwings are common. At the crest, a road runs south about 6 miles to Sounding Knob, an especially good area for Ruffed Grouse, Black-throated Green and Chestnut-sided warblers, and Rose-breasted Grosbeaks.

In the next valley, 19 miles (without detours) from the county line, is Monterey, the county seat, a little town with an inn, a motel, and a couple of restaurants. If you want a map of Highland County, ask the way to the office of the Highway Department.

Take a right at the flashing light onto U.S. 220. From this intersection, drive 5.4 miles to a trout hatchery on the right, worth checking for Wood Ducks, Green-backed Herons, and other migrant water birds. At 6.4 miles turn sharp left on Road 642, which runs beside the South Branch of the Potomac River. A Warbling Vireo nests on the corner.

The circuit described in the following paragraphs is the route to take if you are looking for Golden Eagles in the winter. (Note: For the protection of the eagles, do not mention them to residents of the area.) After 1.2 miles on Road 642, the ridge called the Devil's Backbone rises ahead on your right. The eagles often use it as a roost and circle above it. At 2.4 miles take Road 640, the first right in the village of Bluegrass, and at 3.9 miles go left on Road 644, or continue 2.0 miles to the state line and retrace your steps. Check all the snags and flying raptors. (Incidentally, the roads around Bluegrass are excellent in summer for Cliff Swallows and Savannah and Vesper sparrows, and you may see Red-headed Woodpeckers.)

At 5.5 miles (9.5 if you went to the state line) turn left on Road 643 (or continue a bit on Road 644 and come back), and in 1.0 more mile turn left again on Road 642. At the next junction, in 0.9 mile, turn right on Road 640 and drive at least the 6.3 miles to U.S. 250, about 5 miles west of Monterey. (This is a likely stretch to look for Bobolinks in June.) If you have not seen an eagle yet, continue south on 640 for

4.3 miles, turn left on Va. 84, turn left again after 2.4 miles on U.S. 220, and return to Monterey.

Other raptors that winter in the county include Red-tailed, Red-shouldered, and Rough-legged hawks, American Kestrels, and once, at least, a Bald Eagle.

If you come in early summer, go west on U.S. 250 from the intersection with Road 640, crossing into West Virginia at the crest of Alleghany Mountain. Along the first half-mile from the intersection look for Bobolinks and Cliff Swallows, then for Golden-winged Warblers in appropriate habitat. At 3.4 miles the edges of the stream and the old beaver ponds can be rewarding, and at 6.3 miles Road 601 south makes an interesting side trip for sparrows and Ravens.

In West Virginia, a left turn at the junction with U.S. 28 (14.7 miles) leads in 2.2 miles to Bartow, where there is a motel suitable as a base both for exploring the bird-rich Cheat Mountains beyond and for birding the northwest corner of Highland County. For the latter pursuit go right, instead, on U.S. 28, and after 6.5 miles turn sharp right on a gravel road at the sign for Locust Spring Picnic Ground.

In 0.5 mile the road meets a T-junction. The road right immediately crosses back into Virginia and provides splendid birding as far as you want to go. The road left leads via the first right fork to the picnic ground in 0.6 mile. You are at about 4,000 feet here, and you are in Virginia after going through the open gate.

Nesting birds along the roads and nearby trails include Least Flycatchers; Veeries; Hermit Thrushes; Golden-crowned Kinglets; Solitary Vireos; Black-throated Blue, Black-throated Green, Magnolia, Canada, and Blackburnian warblers; Rose-breasted Grosbeaks; and Purple Finches. Try the path down to the beaver ponds beyond the end of the road into the picnic ground.

The best location for nesting Mourning Warblers is Paddy's Knob. From Monterey go south on U.S. 220 4 miles; then go west on U.S. 84 15 miles to Monongahela National Forest and the Virginia–West Virginia line. Just beyond the sign turn left on unpaved Forest Road 55, drive 2.8 miles (through gorgeous flaming azaleas in spring), and park on the left shoulder where an abandoned road parallel to your route is visible. Put on your waterproof boots for protection from the wet vegetation and walk back along the old trail until the canopy overhead gets thick and the briers on either side disappear. Up to four pairs of Mourning Warblers breed in the brier patches along this stretch, as well as one or two more along the trail ahead of your car. If you should see one on the west side of the road, you can put it on your West Virginia list! This site is also a good place to see the species listed under Locust Spring.

Note on chickadees: The common species in Highland County is the Black-capped, but Carolinas and Black-capped × Carolina hybrids do occur. Identification cannot be deduced from location alone.

16 Washington and Allegany Counties, Maryland

In the heart of Maryland's ridge-and-valley province, Washington and Allegany counties offer a scenic landscape and resident Wild Turkey, Ruffed Grouse, Common Raven, and (west of Hancock) Black-capped Chickadee.

Breeding species include Worm-eating, Golden-winged, and Cerulean warblers. Migration time, spring or fall, can provide unsurpassed land birding: try the first three weeks in May just after a day or two of stormy weather, or an autumn day following the passage of a cold front. Allegany County offers in addition a lake big enough to pull in migrating water birds, a newly discovered hawk watching site, and a good shorebird pond. Some of the very best birding requires you to drive on unpaved back roads. All those mentioned here are navigable in two-wheel-drive vehicles, and you will want to go slowly, for the sake of both protecting your car and spotting the birds.

As soon as you start thinking about visiting Allegany County, you should write or call Green Ridge State Forest, Star Route, Flintstone, MD 21530; (301) 777-2345. Ask to be sent the hiking trail map, the road and campsite map, and the fall-color tour guide. They will enhance your visit considerably. If you have not planned ahead, when you enter the county go to the forest headquarters, a well-marked exit from I-68 17 miles west of the turn-off from I-70. The staff is friendly and helpful, and their maps more accurate and less confusing than the state publications. The only time birders and hunters are likely to coincide in the forest is mid-April to mid-May, the spring turkey hunting season, but local birders do not seem to be deterred by this competition. In late November both turkeys and deer are hunted, but it is not the best time for a birding visit.

Blairs Valley. From the Capital Beltway, I-495, drive north and west via I-270 and I-70 for 69 miles to the exit for Clear Spring. Enter Md. 68 west at mile 0.0 and get in the left lane. Cross U.S. 40 onto Mill Street, where there is a sign for the Indian Springs Wildlife Management Area. At 0.5 mile turn right at the T-junction on Broadfording Road, and at 1.1 miles go left on Blairs Valley Road. At 3.5

miles turn left into a small parking lot just below the dam at the south end of Blairs Valley Lake.

Ravens nest on top of the ridge across the lake, and in the spring turkeys and Ruffed Grouse can be heard calling from the hillsides on both sides of the valley. Barn Swallows nest abundantly under the eaves of the barn near the main parking lot up the road at 3.7 miles, Purple Martins flourish in the martin houses, and the numerous bluebird boxes support expanding populations of Tree Swallows and Eastern Bluebirds. Look for American Bitterns, Soras, and Swamp Sparrows in the marsh below the dam.

If you walk around the south end of the lake, you will find a choice of three trails to take you north on the far side of the valley: close to the lake and the creek that feeds it, near the foot of the ridge, and along the slope. Black-billed Cuckoos; Ruby-throated Hummingbirds; Willow Flycatchers; Warbling Vireos; Golden-winged, Yellow-throated, Prairie, and Yellow warblers; Yellow-breasted Chats; Northern and Orchard orioles; and Blue Grosbeaks (not always easy to find) breed in the valley in the varied streamside, sycamore, and hedgerow habitats.

The most strongly recommended riverside birding in Washington County is **Big Pool** in Fort Frederick State Park. Return to I-95 and continue to the next exit, for Indian Springs. Go east on Md. 56 2.0 miles to the park and follow the signs for 1.0 mile to Big Pool. Park in the lot just above the towpath or in the one by the beaver dam just below it. Because most of the canal is dry, this swampy area and the bottomland woods between the towpath and the river are especially productive. Brown Creepers, Warbling Vireos, and Prothonotary and Yellow-throated warblers nest here, and it is the best place in the county for migrating songbirds.

Return to I-70, continue west for about 6 miles, and take the Hancock exit on Md. 144. Drive west 3.4 miles to the far side of town (keeping left at the fork) and turn left on Round Top Road (mile 0.0).

At 3.3 miles turn left on Willow Road and immediately left again on Sevolt Road. (You could go straight on Willow Road, but Sevolt is the scenic route.) Sevolt Road ends at 5.1 miles, with Orchard Road coming in from the right. Go left on Deneen Road and park at the bend in 0.1 mile. Cross the canal bed on the left and walk east along the towpath.

This stretch is another good spot for Common Raven, Wild Turkey, and Ruffed Grouse; Brown Creeper; Black-capped Chickadee; Warbling Vireo; Cerulean, Redstart, and Parula warblers; and occasionally Prothonotary and Yellow-throated warblers. If you are particularly looking for Ceruleans, here or anywhere, keep in mind that their favored habitat includes an abundance of grapevines.

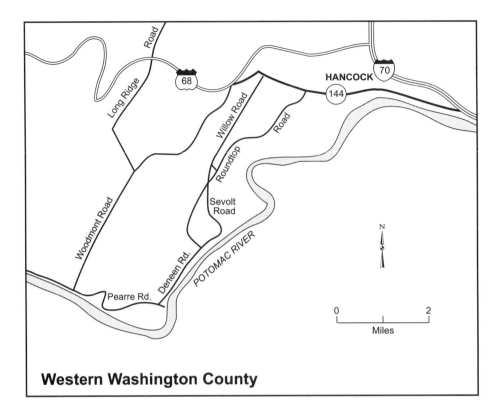

Western Washington County

Deneen Road ends at a T-junction 2.3 miles farther on. Turn right on Pearre Road, which soon becomes unpaved. Look for turkeys and grouse in the woods. In 3.8 miles at a T-junction, turn right and uphill on Woodmont Road. In the pine woods up on the ridge 2.0 to 2.5 miles from the T-junction you may find Black-capped Chickadees and Pine Warblers. In another 4 miles you will reach I-68 just after meeting Md. 144 and turning left to the access road. A right turn (U.S. 40 east) will take you back to I-70 in 3.8 miles. Go left, and you will soon be in Allegany County.

Twelve miles west of I-70 take the Orleans Road exit south. In 3.5 miles go left at the fork and follow High Germany Road south to the Fifteen Mile Creek campground and access to the C&O Canal. (The right fork will also get you there; it is a better road but less interesting.)

From the canal go under the overpass and continue through the tiny community of Little Orleans, a point of access to **Green Ridge State Forest.** Continue straight to the T-junction beyond the ford (passing the terminus of Orleans Road—a direct 5.7 miles from I-68—on the right). Turn left on Oldtown Road to begin a long winding

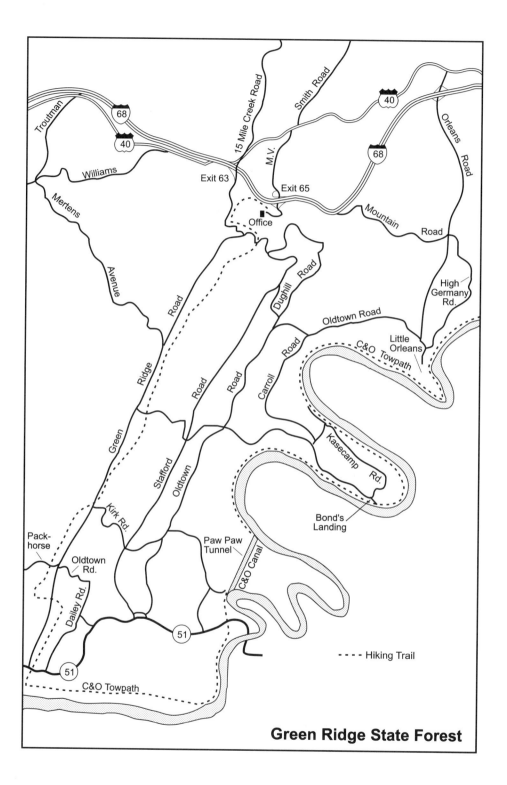

Green Ridge State Forest

- - - - Hiking Trail

tour of the forest that should take you through miles of turkey and grouse habitat, down to the Potomac, and along high ridges. (After heavy rains Oldtown Road is flooded within yards of the T-junction, and you can back up and turn around.)

Drive very slowly, with your window open, so that you can stop wherever you hear the drumming of grouse, the call of a turkey, or the songs of birds you want to see. Most of this route is recommended by the forest staff for its beauty and historical interest, and it abounds in scenic overlooks and good hiking trails. Take your spotting scope along, and look for turkeys on the open slopes below the western overlooks.

At the first fork, where Oldtown Road goes right, keep left on Carroll Road, and stop in a quarter-mile at Point Lookout, 300 feet above the Potomac, for the magnificent view. At the end of Carroll Road take a sharp left by the north end of Stickpile Tunnel onto Kasecamp Road, which follows a loop of the Potomac to Bond's Landing, where you can bird along the towpath. Brown Creepers, Warbling and Yellow-throated vireos, and Prothonotary, Cerulean, Parula, and Yellow-throated warblers may all be here. Just beyond the south end of Stickpile Tunnel turn left on Mertens Avenue, which climbs to Banner's Overlook at the top of Town Hill (1,650 feet above sea level). A hike in early summer along Stafford Road, which runs along the Town Hill ridge, may produce a surprising number of Cerulean Warblers.

Continue west on Mertens Avenue, which drops into East Valley and then rises to Green Ridge. Turn left on Green Ridge Road. From this point you can drive 4 miles south out of the forest onto Md. 51, which runs along the southern edge of the county, or you can drive down and back. In addition to the grouse, turkeys, and Black-capped Chickadees, Woodcock, Solitary Vireos, and Pine and Hooded warblers are among the regular nesters along the road. If you are looking for Golden-winged Warblers, try the corner of Green Ridge Road and Kirk Road. Walk a short distance along the hiking trail that runs north from this intersection. If you have no luck there, go down to Oldtown Road, almost across from the junction of Green Ridge Road and Packhorse Road. Oldtown leads to an old orchard where the warblers are usually found.

If you are interested in more birding along the canal, turn right on Md. 51. The points of access comprise Oldtown, 5 miles to the west; Spring Gap, 7 miles farther on; and North Branch, 2.5 miles beyond that.

For Allegany County's good shorebird spot, continue past the North Branch parking area a half-mile to a fenced area around a smelly soybean processing plant. Pass the road with the yellow gate and the "No Trespassing" sign, and take the narrow dirt road to the left. (You may want to park on the road and walk in.) Two sewage lagoons are behind the fence; beyond them, outside the fence and beyond it, is a large pond with both water and mud flats. You can see much of it from the end of the road, but if shorebirds are there you will probably want to take the trail over the mound at the end of the road. It leads to another road on the far bank of the pond. Afternoon light is best, and you will want a scope. Though dabbling ducks are frequent visitors and gulls and American Pipits drop in, it is the possibility of seeing interesting shorebirds, mainly in fall, that gives this site its reputation. Birds on the list include American Avocet, Lesser Golden-Plover, and Baird's Sandpiper.

If it is spring and you are spending the night around Cumberland, turn left on a loop road through an industrial park when you leave North Branch instead of returning to Md. 51. The undeveloped stretches are outstanding for displaying woodcock at dusk and dawn. You will leave the area by way of a right turn on Mexico Farms Road, which runs into Md. 51.

In 4.7 miles, on the outskirts of Cumberland, turn right on Wineow Street, go under a bridge, cross a railroad track, and follow the road up onto the towpath at the western terminus of the C&O Canal. This can be a good area for Great Blue and Green-backed herons, Black-crowned Night-Herons, dabbling ducks, Osprey, and Killdeer. If the water levels are low, other shorebirds, Ring-billed and Bonaparte's gulls, and Forster's, Common, and Black terns are found in migration, especially in May.

If you made a U-turn on Green Ridge Road and headed north again instead of leaving the ridge, a left turn at the end onto Fifteen Mile Creek Road will take you back to I-68. Continue across the highway onto Scenic U.S. 40 eastbound. In 4.3 miles you will be at the top of **Town Hill.** The Town Hill Inn is on the left and a large grassy area is on the right. A long, open-sided, roofed shelter is between you and the gorgeous view of the valley to the east. This is the new hawk-watching site, first staffed in 1990. The documented report of a Black-shouldered Kite and 20 Golden Eagles—8 of them in one day—in the first season is attention-grabbing. Counts of 15 other species of raptors, some in good numbers, as well as 119 loons over the fall season and 9 ravens all at once should be enough to lure you up there some October day. To reach the site from the east, take Scenic U.S. 40 3 miles after crossing the county line, or join it from I-68 at the Orleans Road exit.

The other site worth a visit, at least during waterfowl migration, is **Rocky Gap State Park,** 13 miles west of the State Forest Headquarters. There are numerous vantage points from which to scan the lake, from the Touch of Nature Trail at the western end to the campground entrance road at the eastern end, 1.5 miles up Pleasant Valley Road.

Take advantage of them all. You have a good chance to see (in the course of a year) Common Loons, Horned and Pied-billed grebes, perhaps a Red-necked Grebe, Green-winged and Blue-winged teal, American Wigeon, Ring-necked Ducks, both scaup, Canvasbacks and Redheads, scoters and mergansers, Ring-billed and Bonaparte's gulls, and Black Terns.

If you drive to the campground when you first arrive, turning right beyond the park office, you may see turkeys around the picnic pavilions on the left or up on the slopes on the right. The open fields sometimes attract Horned Larks and pipits, and the woods at the western end of the lake may shelter Golden-crowned Kinglets and a variety of migrating warblers as well as the nesting Pine Warblers.

17 The Appalachian Plateau

When the flood of spring transients has slowed to a trickle at the end of May, think seriously about heading out for a long weekend in western Maryland, where Bobolinks dance over the fields, Alder Flycatchers sing in the swamps, and Saw-whet Owls pipe in the night.

The mountains of Garrett and Allegany counties are so laced with quiet back roads, state parks, wildlife areas, and upland farms that any birder with a drop of explorer blood can find a dozen private birding spots in a few days of investigation; the sites described here are just a few of the essential ornithological landmarks.

The straightforward nonstop route from Washington is I-270 to Frederick, I-70 west to Hancock, and I-68 through Cumberland to Exit 29 for your first stop. To reach **Finzel Swamp,** an area renowned for its nesting Black-billed Cuckoos, Alder Flycatchers, Golden-winged Warblers, Northern Waterthrushes, Rose-breasted Grosbeaks, Swamp Sparrows, Purple Finches, and sometimes Henslow's Sparrows, go north on Road 546 for 1.5 miles and turn right onto a dirt road just beyond a baseball field marked by the sign "Eastern Garrett County Recreation Area." Follow the dirt road 0.1 mile to a quadruple fork. Take the second branch from the left and continue 0.5 mile, passing swampland on your left, to the parking area in front of the gate at the end of the road. Walk through the gate and follow the path through open meadows to a pond about 0.25 mile beyond.

Back on I-68, continue to U.S. 219, Exit 14A, and go south 26 miles across Deep Creek Lake to Oakland, which is, for birders, the most central town with good motels in Garrett County. To reach the main birding areas go west on Green Street, a right turn (from the north) at

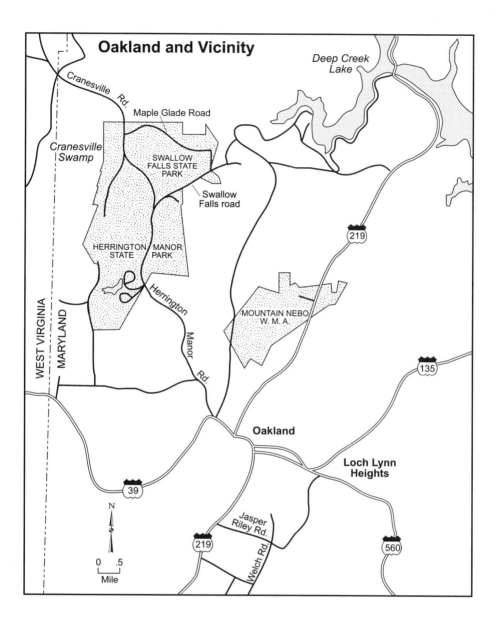

Oakland and Vicinity

Cranesville Rd.

Maple Glade Road

Deep Creek Lake

Cranesville Swamp

SWALLOW FALLS STATE PARK

Swallow Falls road

HERRINGTON MANOR STATE PARK

Herrington Manor Rd.

MOUNTAIN NEBO W. M. A.

WEST VIRGINIA

MARYLAND

Oakland

Loch Lynn Heights

Jasper Riley Rd.

Welch Rd.

N

0   .5
Mile

the Garrett National Bank sign. It soon turns into Herrington Manor Road. The road runs northwest 4.2 miles to **Herrington Manor State Park** and continues another 2.0 miles to a fork. Bear left on Cranesville Road for Cranesville Swamp; bear right on Swallow Falls Road for Swallow Falls State Park.

Try to visit Herrington Manor as early in the morning as you can get in; if the entrance gate is closed, bird along the entrance road or poke along the highway. Look for nesting Golden-crowned Kinglets in the conifers. (In these trees the first nesting record of Pine Siskins for Maryland was confirmed.) Once inside, work your way around the artificial lake in either direction, not wasting much time in the red-pine plantation but exploring thoroughly the deciduous trees, the alder swamp, and the white-pine forest off to the left of the visitor center. Red-headed Woodpeckers; Alder Flycatchers; Veeries; Hermit Thrushes; Golden-winged, Chestnut-sided, Magnolia, and Black-throated Blue warblers; and Rose-breasted Grosbeaks are among the summer residents. The chickadees you see and hear will be Black-capped.

The visitor center at Herrington Manor serves both this park and Swallow Falls. Drop in and pick up maps to the roads and hiking trails.

The main entrance to **Swallow Falls State Park** is 3.2 miles north; on a holiday weekend especially, it pays to beat the crowd, as some of the most interesting birds are close to the parking lot, in the hemlocks along the river that nonbirders enjoy just as much as you do. Golden-crowned Kinglets nest in the spruces around the parking lots and picnic areas. Look for Solitary Vireos and Black-throated Green, Blackburnian, Canada, and Magnolia warblers in the forest and Northern Waterthrushes on and over the river rocks. Beyond the parking lot, Maple Glade Road—unpaved but pleasant—departs northwest past park headquarters and ends at Cranesville Road in 2.0 miles. The narrow road gets little traffic and is especially good for Least Flycatchers. Park carefully and explore the old trails that branch off on either side for a better look at some of the species that like a mixed coniferous-deciduous forest. Birders who have thoroughly explored the plateau consider this road one of the best stretches for birding in the region, especially the boggy, scrubby areas in the second mile.

At Cranesville Road turn right and drive 2.3 miles. The road eventually descends toward **Cranesville Swamp** and winds along its eastern edge for the next three miles. Take the second left onto Lakeford Road at the sign for "Cranesville Swamp Nature Preserve." Drive 0.5 mile, keeping straight at the fork, and turn right at the Cranesville Swamp sign. Continue to the parking lot on the right. Follow the Nature Conservancy sign instructions to the boardwalk through the swamp as well as to the loop trail. (Those who keep state lists should be aware that this part of the swamp is in West Virginia.)

Make a night visit to hear the Saw-whet Owls, a challenge to sort out from the hundreds of noisy frogs. Come back in the morning to study the songs of Alder and Willow flycatchers, which both nest in the alders, the former apparently to the south of the square house at Muddy Creek Road (0.1 mile south of the entrance) after the bog widens out and is broken up with stands of young conifers, the latter mostly to the north where the bog is narrow and open. (Both species have been found around the church 0.1 mile farther up Cranesville Swamp Road.) Other inhabitants include Golden-crowned Kinglets, Nashville Warblers, Swamp Sparrows, and many of the species listed under the state parks.

A small swamp close to Oakland is reached by taking U.S. 219 north from Md. 135 about 3.5 miles and turning left into the **Mount Nebo Wildlife Management Area.** Park at the gate and walk about a half-mile to the swamp, which has at least one pair of Alder Flycatchers, plus Veeries, Golden-winged and Chestnut-sided warblers, and Swamp Sparrows. The first left beyond the swamp leads to a beaver dam and ten acres of open water where ducks and geese nest.

The best area to find nesting Mourning Warblers in recent years has been on **Backbone Mountain** off Table Rock Road. Drive south from Oakland on U.S. 219 for 8 miles to U.S. 50, and turn left (east). In 2.3 miles turn right on Table Rock Road, drive 0.9 mile and take the right fork onto the gravel road. Go 0.1 mile and park at the locked gate. Ignore the "No Trespassing" sign; birders are welcome. Walk up the road toward the Roth Rock fire tower. Halfway up, the road bends to the right and enters the woods. Look for the Mourning Warblers among the downed trees and brier patches in the intermittent clearcut areas on the left for the next half-mile or so. Other nesting species on the mountain include Chestnut-sided and Hooded warblers, Rose-breasted Grosbeaks, and Dark-eyed Juncos.

When bird song dies down in the afternoon, use the time in exploration of habitat. A likely stretch is reached by taking U.S. 219 south of Oakland and then looping back on several of the roads through farmland to the left and right of the highway. The first left, Jasper Riley Road, goes past Welch Road, a spur 1.1 miles to the left that leads north to a fenced-off waste water treatment plant. Savannah Sparrows and Bobolinks may be on both sides of Welch Road, and in the boggy meadow on the western side the sharp-eyed may find Upland Sandpipers, while Willow Flycatchers sing beside the little stream.

Try any and all of the other side roads on both sides of U.S. 219 as well. Several of the farms in this Amish community have feeders, often with interesting visitors that you can see easily from your car, and both the fields and stream crossings are worth checking.

Though few birders from the Washington area visit Garrett County before the breeding birds arrive, the concentrations of loons, grebes, and waterfowl on **Deep Creek Lake** can be spectacular in migration (mainly November and April). The view of the lake is obscured in many places by weekend cottages, but an efficient tour of the northern half of the lake can be made by following this route: from Oakland, take U.S. 219 north to Glendale Road, turn right, cross Glendale Bridge, and park at the east end. You can scope much of the lake from here. Then take the next left onto State Park Road and take every road that follows the shoreline until you come back to U.S. 219. There are many vantage points.

The Western Shore

Eastern Baltimore County and Conowingo Dam 18

The proximity of Baltimore to Chesapeake Bay provides the birders of that city easy access to much-wanted birds that rarely, if ever, turn up close to Washington.

All stops on this tour are on the far side of Baltimore Harbor. To reach them you must pay $1 each way to take one of the tunnels or the Francis Scott Key Bridge (or squander the equivalent in gasoline by driving all around the Baltimore Beltway, I-695).

If you want to make a full day of it, rise early and head for **Gunpowder Falls State Park,** renowned for resident owls, wintering waterfowl and sparrows, and breeding Least Bitterns, Virginia Rails, Whip-poor-wills, Bald Eagles, and a wide assortment of passerines.

The most direct route is by way of the Fort McHenry Tunnel and I-95 to Exit 67A, Md. 43 east, 3.1 miles north of I-695. Go east 1.4 miles to U.S. 40, then north 0.6 mile to Ebenezer Road, Md. 149, and follow it 4.2 miles east to the park; Ebenezer Road becomes Grace's Quarter Road in the village of Chase.

The main entrance to the park, the Hammerman Area, opens at 8 A.M. from Memorial Day to Labor Day and at 10 A.M. off season. A $5 fee is charged, with an extra $1 charge for out-of-state visitors, except from October to April or on weekdays in May and September.

An always-free section lies farther down the road, which forks soon after the right turn for the marina. Bear right and park 1.3 miles from the park entrance. Walk down the track that forks left from the continuation of the road. In spring and summer, Virginia Rails in the small marsh respond readily to a tape; Yellow-breasted Chats and Orchard Orioles sing around the edges.

The track ends at Dundee Creek, with winter views of Tundra Swans, American Wigeon, and other dabbling and diving ducks. You can walk to the right along the shore to a boat ramp and back 0.1 mile along the road to your car.

The stand of pines on the road above the parking area should be checked for owls, Eastern Bluebirds, and Red-breasted Nuthatches. Any trail can be good for land birds at any season.

The main park is well worth a visit, especially out of season. Park at Picnic Area A and walk down to the marsh, where Virginia Rails and Least Bitterns are present all summer. Then drive down to the boat ramp and check the trees across the creek that serve as a Bald Eagle roost.

◀ Forster's Terns (Courtesy Felix C. Lowe)

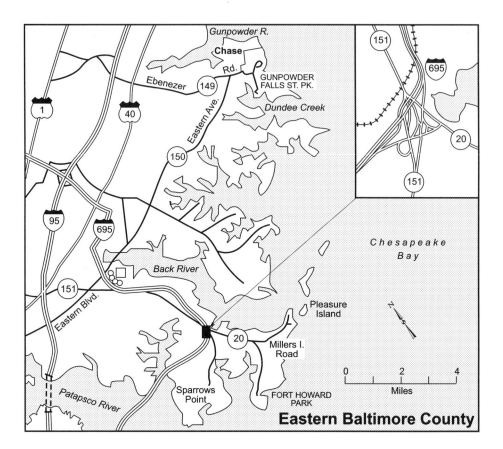

Eastern Baltimore County

The Gunpowder River, overlooked by the beach, is excellent for waterfowl in winter, especially for Common Mergansers. On the far side of the parking lot a trail leads out into yet another marsh.

The center of the park is being allowed to revert from lawn to abandoned field. It should be worth checking for sparrows in any season.

On leaving the park, go back 0.5 mile to Grace's Quarter Road and turn left. Go 0.1 mile to Eastern Avenue, Md. 150, turn left, and follow it all the way back to Baltimore. (It is variously named Eastern Avenue and Eastern Boulevard.) After about 8 miles, the road crosses Back River.

At the next light go right on **Diamond Point Road.** In 0.2 mile, turn left into the shopping center parking lot and park. Walk back to the road and cross over to the hard grassy fields that go out to the river's edge. Except at full high tide, extensive mud flats lie before you, very popular with gulls and terns though not much used by shorebirds. In late August and September it is the best location in the region to look for Franklin's Gulls among the thousands of resting Laughing Gulls. The light is always good.

Drive back to Eastern Boulevard and cross it straight into the **Back River Waste Water Treatment Plant,** perhaps the best birding area in Baltimore. You are welcome to drive or walk anywhere.

(If you want to skip either Gunpowder Falls State Park or Miller's Island Road, the direct route to or from the Back River Waste Water Treatment Plant is via Francis Scott Key Bridge and I-695 to Md. 150, exiting east to the first light. Go right for the plant, left for Diamond Point Road.)

Check in with the guard at the entrance gate. Then enter the visitor parking lot immediately on your left. At the far end of the lot are two small ponds and a nature trail. This area has abundant shrubby growth and wooded wetland habitat. It is an excellent area for various passerines.

The adjacent pine stand near the entrance was famous one winter for a visiting Boreal Chickadee; it is always worth checking for owls, Red-breasted Nuthatches, and migrants.

Return to the entrance road (Willis Avenue) and turn left. Check the four round settling tanks on your left and then continue straight ahead onto Primary Way. Here, check the seven round settling tanks. In the fall, look for Franklin's Gulls among the Laughing Gulls, mostly sitting on the grass. In the winter, Ring-billed Gulls fly around and over the tanks; they are sometimes joined by a Common Black-headed Gull. In March and April, Bonaparte's Gulls predominate, with occasional Little and Black-headed gulls among them.

Retrace your route to the first right turn (Willis Avenue), which will take you downhill past the stone wall of the abandoned trickling filters.

Where the paved road curves left, keep straight on the gravel road and drive as far as the entrance to an abandoned landfill to check the cattail and mallow marsh on the right for herons, shorebirds, and rails.

Then go back to the corner, turn right at Willis Avenue, and right again at the next intersection, Activated Avenue. The road ends at Back River. The mouth of Bread-and-Cheese Creek is on your right and is especially good for ducks, small gulls, and shorebirds. The shallow pools along the railroad track by the creek can be excellent in shorebird migration for Black-bellied and Lesser Golden-Plovers; Pectoral, Spotted, and Solitary sandpipers; both yellowlegs; and all the "peeps."

Retrace your route on Activated Avenue and go about twenty yards beyond the intersection with Willis Avenue. Turn right on an unnamed road that is bordered by twelve final-clarifier tanks. These tanks can be excellent in the appropriate season for the various gulls men-

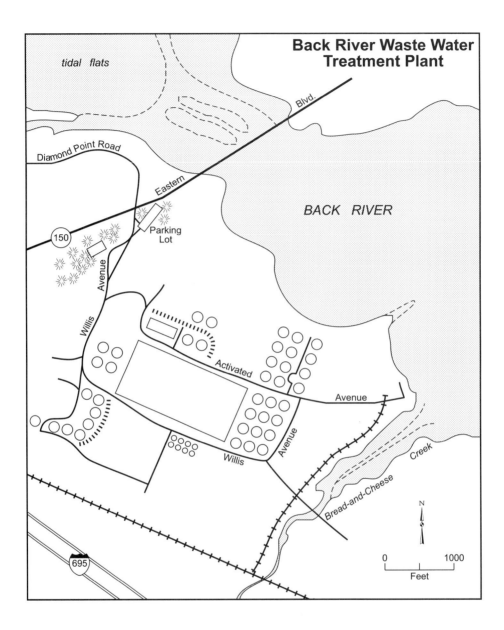

Back River Waste Water Treatment Plant

tidal flats

Diamond Point Road

Eastern

Blvd.

150

Parking Lot

Avenue

Willis

BACK RIVER

Activated

Avenue

Avenue

Willis

Bread-and-Cheese

Creek

695

N

0   1000
Feet

tioned earlier. (This was the most reliable spot in March and April of 1990 to find the Ross's Gull.)

Explore the north side of the plant if you have time. There are sparrowy fields and swampy woods. Blue Grosbeaks are common in summer and a Cooper's Hawk is resident in winter. Red-tailed and Red-shouldered Hawks often circle overhead.

When you leave, turn left and after 0.5 mile head toward Sparrow's Point on I-695, just west of the plant. After 4.8 miles, keep right on

Md. 151 south, get in the lane for Edgemere and Fort Howard, and exit at 5.3 miles, keeping in the right lane for Md. 20. At the T-junction at 5.8 miles, turn right on Md. 20. Continue south about 1.5 miles and turn left on **Miller's Island Road.** Measuring from the turn, after 0.7 mile, note Hertzinger Road on the left. On spring evenings Woodcock use the field west of this road as a display ground, and both Whip-poor-wills and Chuck-will's-widows nest in the adjacent woods. You can pull into Hertzinger Road to watch and listen (but note that there is room for only one car in front of the locked gate); then continue on Miller's Island Road.

As parking is illegal at the following three sites for rails, note their locations as you drive by. You can park at 1.3 miles, the first legal area to do so, and walk back. On the right (as you drive) at 1.0 mile from Md. 20 is a vast marsh with nesting King and Virginia rails, migrant Soras, Common Moorhens, and American Bitterns, and breeding Least Bit-terns, Marsh Wrens, and Swamp Sparrows. In the warmer months all the local herons (except Yellow-crowned Night Herons), egrets, and ibis can be seen. Ospreys hunt in summer and Northern Harriers from fall to spring, and wintering Common Snipe fly in and out.

Another good spot is at 1.1 miles on the right. Rails are not as close to the road here, but the very best site for Soras is at 1.2 miles. At 1.3 miles, park at the junction with 12th Street on the left and walk back to these sites.

Upon returning to your car, walk the block and a half to the end of 12th Street, where the owner of Augie's Carryout Crabs regularly feeds and attracts numerous waterfowl. In the winter this is one of the better spots to see Canvasbacks, Redheads, and scaup.

Continue east on Miller's Island Road. At 0.1 mile continue straight onto Cuckold Point Road. At 0.4 mile turn right at the entrance to Fisherman's Inn. Park near the road, not near the restaurant. Walk with your scope out to the water.

Across the channel in front of you is Pleasure Island, which has produced a large shorebird list in late July: Piping Plovers, Red Knots, Whimbrel, Ruddy Turnstones, and Sanderlings are all among the Western Shore rarities. Skimmers may appear in late August, and large rafts of scaup and flocks of Bonaparte's Gulls in spring.

The pilings back to your left are worth checking for terns: For-ster's are the most common, but there are noteworthy records of Caspian, Royal, and Black Terns as well. All the rarer gulls have been seen here, including a Little Gull in summer, and loons, grebes (sometimes a Red-necked Grebe), and waterfowl are reg-ular winter visitors.

Return to Md. 20 (North Point Road) and turn left. The fields south of Miller's Island Road should be checked for shorebirds and for Franklin's Gulls among the Laughing Gulls.

In 0.6 mile turn left at Bay Shore Drive and enter Black Marsh State Park. The park is open Wednesday through Sunday, from 8 A.M. to 4 P.M. Park in the parking lot. It is a 1.5-mile walk along the old roadbed to the water to look for the Bald Eagles that currently nest in the park. If the park is fully developed for recreation, as announced in the spring of 1991, this stop is unlikely to remain worthwhile.

Fort Howard Park, 2 miles farther south, is worth a visit in migration, especially for hawk watching. The park is open year-round from 8 A.M. to dark.

The last stop is on the way back to Washington. At the service station on Md. 20, 1.1 miles north of Miller's Island Road, go left on Sparrow's Point Road. Turn left in 1.0 mile on Md. 151, turn right in 0.3 mile on Wharf Road, bear right in 0.2 mile toward Francis Scott Key Bridge and Dundalk, and immediately park on the shoulder for a quick check of the pond below you on the right. Common Moorhens, King Rails, and Wood Ducks have nested here; wintering ducks include American Wigeon, Shoveler, Gadwall, Hooded Merganser, Bufflehead, and Common Goldeneye. Don't step over the railing or take photographs. At the bottom of the hill, scope the large pond on the right for waterfowl and gulls. Then merge onto Bethlehem Boulevard and follow the signs to Francis Scott Key Bridge.

If you want to plan your whole day around gull watching, you cannot do better than to combine a visit to Back River with one to **Conowingo Dam.** This site, where U.S. 1 crosses the Susquehanna River precisely on the eastern edge of the piedmont, is a point of pilgrimage for gull enthusiasts from throughout the region. Except for one year when there was a serious fish kill upstream, the rocks, water, and air below the dam consistently have had the highest concentration of gulls on or near Chesapeake Bay. In addition, the dam provides the unusual opportunity to look down at gulls in flight and gain valuable experience in learning the variability within common species.

To reach the dam take I-95 62 miles north from the Capital Beltway. At Exit 89 go west on Md. 155 2.8 miles to Md. 161. Turn right and continue 5.4 miles to U.S. 1. Turn right. In 1.6 miles turn right onto Shuresville Road at the sign for Conowingo Fisherman's Park, go 0.6 mile to Shure's Landing Road, and turn sharply left. Drive down to the river below the dam. You can park and scan the river at the foot of the hill and then continue to the parking lot just below the dam, where you will find rest rooms and a picnic shelter.

From December to about mid-February the extensive rocks on the far side of the river are covered with gulls. Other birds regularly present include large numbers of Great Blue Herons and Black-crowned Night-Herons, Turkey Vultures, and Bald Eagles, with a Golden Eagle sometimes among them. Common Mergansers winter on the river above and below the dam.

A spotting scope is required if you are to enjoy Conowingo, even when you walk out along the long platform under the dam and are closer to the gulls. The best days to be there are the ones when there is extra demand for electricity, and the generators are running much of the time. (This situation rarely occurs on Sunday, and happens intermittently on Saturdays.) When the generators are turned on, bells ring, lights flash (to warn the fishermen on the shoreline that the water is about to rise), water (with fish, or fish pieces, in it) gushes from the dam, and all the gulls surge into the air to feed in the churning waters.

Among the thousands of Herring and Ring-billed gulls and hundreds of Great Black-backed Gulls always present, you are likely to find a few Lesser Black-backed Gulls, frequently an Iceland Gull or two, and sometimes a Glaucous or a Thayer's gull. Laughing Gulls and (rarely) a Franklin's may be around in September, and Bonaparte's Gulls are present in early December, perhaps a good time to look for Little or Common Black-headed gulls. Few birders, however, visit the dam before Thanksgiving or even Christmas. A January visit usually promises the greatest numbers and variety.

Baltimore Harbor 19

Baltimore Harbor offers close-to-home opportunities for watching grebes, cormorants, herons, ducks, shorebirds, and gulls, as well as raptors and passerines, all in their appropriate seasons. Rarities that have been found here in the last few years include Red-necked and Eared grebes; Glaucous, Iceland, Thayer's, Lesser Black-backed, and Common Black-headed gulls; and a Snowy Owl.

This tour has both scenic and historical attractions. It is limited to the south side of the harbor, requiring no tunnel or bridge tolls for birders coming from Washington.

The harbor shoreline changes constantly, with abandoned land suddenly being converted into a dock, a terminal, or a factory. Most of the birding areas described here should be accessible for some years to come. If you have time, explore the byways in between on your own.

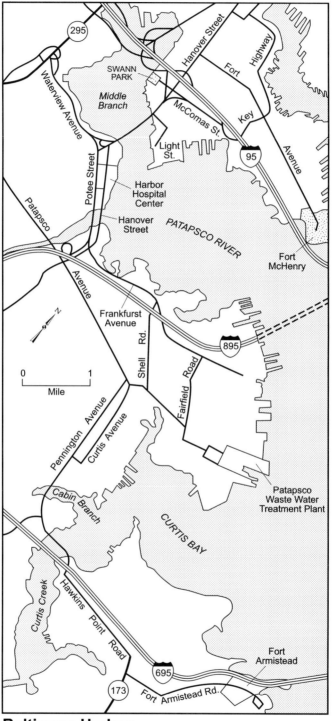

Baltimore Harbor

Take either I-95 or Md. 295, the Washington-Baltimore Express-way, to their shared exit, #54, in downtown Baltimore for Hanover Street, Md. 2 south. Stay in the right lane, bear right into the left-turn lane, and turn left, crossing Hanover at the traffic light onto Cromwell Street.

When Cromwell Street ends in two blocks, turn right on Light Street and drive about a mile to its end at Ferry Bar Park. After looking into the harbor there, walk to your left along the shoreline 0.2 mile to an open area on the right. Walk over to the water's edge on the right, taking care not to flush the hundreds of Canvasbacks, Lesser Scaup, and Ruddy Ducks that may be sheltered by the long pier before you. Look for Redheads among them. (This spot may not be worth check-ing on weekdays, because of human disturbances.)

Go back to Hanover Street and turn right. In 0.7 mile, turn right on West Fort Avenue and follow it 1.8 miles to Fort McHenry National Monument.

Despite its manicured appearance, Fort McHenry has a good reputation as a migrant trap, with such varied visitors as falcons, American Pipits, and Upland Sandpipers, as well as passerines hiding in the trees and bushes. You may find loons, grebes, and Oldsquaw offshore in winter. Wintering Lesser Black-backed Gulls are often found on the channel markers or the pilings in the harbor or along the shore in the cove just to the right of the entrance. (Walk down to the water's edge behind the huge statue of Orpheus.)

Now return to Hanover Street, turn left, and, just beyond the second overpass, go right on West McComas Street. It ends just ahead at Swann Park, a good stop only in the morning. Cross the open field in front of you with your scope, and scan the harbor and the far shoreline for gulls, shorebirds, and herons.

Back at Hanover Street, turn right and follow it across the bridge over Middle Branch, moving to the left lane. On the far side, the highway divides and becomes Potee Street southbound. Turn left at the third traffic light onto Reedbird Avenue and in one block cross Hanover Street into the grounds of Harbor Hospital Center. The parking lot overlooks the harbor in a good area for loons, gulls, the same ducks you saw near Ferry Bar Park, and, when the reservoirs freeze over, Common Mergansers. Drive up to the north parking lot and little Broening Park just beyond it, the site of the Maryland Vietnam Veterans Memorial. A mud flat that emerges across the water at low tide is particularly good for gulls, and you may see two of the city's four resident Peregrines hunting in the neighborhood.

From the north exit, cross Hanover Street on Waterview Avenue and turn left on Potee Street again. Continue south 0.6 mile and fork left on Frankfurst Avenue. At 0.8 mile from the fork, make a U-turn via the left-turn lane for the closed rear entrance to the Maryland Port

Administration and go back 0.1 mile. Just beyond the end of the guardrail, park on the shoulder. Walk down the very trashy fisherman's path to a cove in the harbor. Head straight away from the road; don't follow the more conspicuous path to the right.

This cove is perhaps the birdiest in the harbor, with plenty of ducks, plus the possibility of coots and grebes, and gulls and herons standing along the shoreline and on the old wrecks beyond. The light is always good. Don't be deterred from stopping by the seediness of the environs.

Make another U-turn as soon as you can (0.2 mile). When Frankfurst Avenue forks left in 0.5 mile, keep right on Shell Road, which ends at Patapsco Avenue in another 0.7 mile.

To reach the Patapsco Waste Water Treatment Plant, go left on Patapsco Avenue. In 0.5 mile turn right on Fairfield Road, which becomes Northbridge Avenue. Bear left at the stop sign and turn left on Asiatic Avenue at 1.1 miles. Turn right into the plant entrance in 0.1 mile and ask permission to bird. (It has always been granted, so far.) Check the settling tanks and drive down to the harbor, where it is easy to scan for water birds.

Return to Patapsco Avenue, keeping in the left lane, and, 0.2 mile after you pass Shell Road, turn left on well-marked Pennington Avenue. In 1.2 miles you cross Cabin Branch; if there are birds on the flats on the right or on the pilings on the left, you can park just ahead on the right, where the curb ends. Glaucous, Iceland, Thayer's, and Black-headed gulls have all turned up here, but a nearby landfill has closed since these sightings were made.

The road beyond becomes Hawkins Point Road, then Fort Smallwood Road. At 1.9 miles from Cabin Branch, bear right into the left-turn lane for Fort Armistead Road.

Before you reach the fort, stop 1.3 miles from Fort Smallwood Road, park, and walk across to the water under the Francis Scott Key Bridge. Look for the two Peregrines that began nesting under the bridge in 1989. Check the pilings and wrecks in the cove for white-winged gulls and cormorants (in memory of the Great Cormorant that was seen here once) and the water for Red-breasted Mergansers and scoters.

Continue on Fort Armistead Road to its end in 1.5 miles at Fort Armistead, keeping alert for the trucks that hurtle recklessly out of the paint factory entrance road on the right.

Fort Armistead, a small park with good warbler habitat (try the top of the ruins) and a long fishing pier, looks across to Fort Carroll, a harbor island, and Sparrows Point. The gull-filled pilings beyond the paint company dock on the right are best seen from the end of the fishing pier. Bonaparte's Gulls and various terns are often present in spring.

To reach a westbound access point onto the Baltimore Beltway (I-695), return to Fort Smallwood Road and turn right. After 0.4 mile, turn right onto Quarantine Road. In 0.2 mile turn left and you will be on the beltway.

Sandy Point State Park 20

Birders who seek the varied pleasures of owling, watching spring and fall hawk flights, sorting through gull flocks, building a fat warbler list, scanning for waterfowl, and chasing rarities should sample Sandy Point State Park. At the west end of the Chesapeake Bay Bridge, it lies but 25 miles from the Capital Beltway (I-95/495) just off U.S. 50.

You are unlikely to be able to explore all its diverse habitats thoroughly in one day, but just part of a morning can, if you are lucky, provide birding as rewarding as that to be found anywhere in Maryland. Ask those who saw a Northern Shrike, a Fulvous Whistling-Duck, and a Common Ground Dove one November day, or the ones who viewed a Sabine's Gull before breakfast and had studied a Franklin's Gull and an Iceland Gull by lunch, all on a May morning.

The hours and fees of the park are somewhat unpredictable. At this writing the former range from 8 A.M. to 5 P.M. in winter to twenty-four hours in summer, with a variable expansion in spring and contraction in fall. If you want access at an early or late hour, it would be a good idea to check in advance. The park telephone number is (301) 974-1249.

As you approach Sandy Point, take note of the junction with Whitehall Road on the right. (We will come back to this landmark later.) The approach to the park is by way of an exit to the right, at the sign for Sandy Point State Park. The overpass will deposit you on the left side of U.S. 50, at a fork. The road straight ahead takes you to the park office. The one on the right leads to the tollbooth, which sometimes demands quarters except in the coldest months. (Access is free in winter and on Wednesdays all year.) If you take the first left beyond the tollbooth, you will come to the East Beach parking lot, the best lot to park in for land birding. If you go straight on, you will come to an enormous paved lot on the right serving the marina and the South Beach. Beyond are smaller unpaved lots on the left and at the far end, closed until after the last frost. Park only in a parking lot at all times of the year.

Across the main road from the East Beach lot, a gravel road leads to a small picnic area (where you can park in the warmer months) and

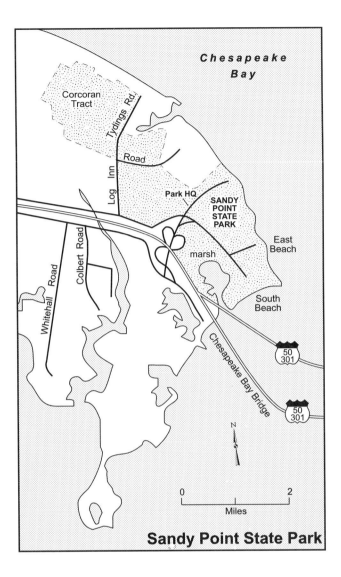

Sandy Point State Park

then to a marsh. There are short foot trails to the left and right; both offer good marsh-edge birding. The latter loops back eventually to the picnic area past a large stand of pines much favored by owls, notably Great Horned, Barred, and occasionally Long-eared. The marsh has year-round Marsh Wrens and Virginia Rails, and all the other rails on the Maryland list have been seen here at least once. The reeds have been taking over the best rail habitat, however.

When you return to the main road, walk back toward the highway from the tollbooth and take the unpaved service road on the right, which will take you through excellent sparrow habitat. At the fork, the trail to the left leads to the road past the park office. The right fork goes through second-growth woods that may be alive with warblers in migration; it ends at the bay beyond a grove of hollies frequented by Cedar Waxwings and Hermit Thrushes. After scanning the bay for waterfowl, gulls, and terns, according to season, walk left along the shore until you reach a set of buildings that used to house park headquarters. (These may be razed eventually.) Follow the paved road inland; it is the one that runs by the current park office. Two dirt tracks on the left form the borders to a pine stand worth combing for owls. The first track merges with the second one, which takes you back to the service road.

If you keep on the paved road as far as the second track, you will come to the back of a sign for the youth group camp. Turning right instead of left, you will be on a road that leads past a tree-lined fence on the right. Through it you can peer at the ducks and gulls at a pond adjacent to a county water treatment plant. There is (at this writing) at least one point where access to the fence is easy, a swath of grass where the cat brier has been eliminated. The road continues through the group camp and on and on through interesting habitat, especially attractive to Blue Grosbeaks in summer. Return the way you came.

If you are mainly after water birds, drive to the South Beach lots and park as close to the beach as you can. The wide beach is a loafing ground for the huge flocks of gulls (up to ten thousand birds) that accumulate here in late April and May. (In winter they prefer offshore ice when there is some.) The buildings between the south parking lot and the beach provide good vantage points and windbreaks. They are rendezvous points for the patient birders who scope with remarkable success for Glaucous, Iceland, Lesser Black-backed, and Franklin's gulls among the hordes of more common species resting on the sand. Bonaparte's Gulls most often occur erratically in March, April, and October, Little Gulls rarely among them, and the only Common Black-headed Gulls that have been recorded here so far showed up one January. These species tend to be seen only in flight or on the water. Rarities may show up at any time from July to May, but you will have to get to the park early to see them when swimming is in season.

All the tern species (except Sandwich) that breed along the Delmarva coast are regular or occasional visitors in the warmer months, as are Black Skimmers.

The long jetty at the far corner of the beach by the Bay Bridge is easy to walk on and offers close views of the diving ducks that winter here, especially Canvasback and Greater and Lesser scaup, and the best chance to see Purple Sandpipers in November. The channel beside the jetty leads into the park harbor; these protected waters can be surveyed first from the jetty, then from a dirt road, and then from the paved parking lot.

You can also walk northeast along the beach, where other, shorter jetties also shelter ducks and gulls and provide habitat for a Purple Sandpiper or two. They are sometimes used by the Snow Buntings that wander the gravel parking lots and the grassier parts of the beach in late fall (often in the company of Lapland Long-spurs).

The man-made hill on the left, with an observation tower on top, is a fine perch for hawk watching. In spring there are major south-to-north flights up the western shore of the bay, especially of American Kestrels, Sharp-shinned Hawks, Ospreys, and Northern Harriers. In autumn, when buteo flights can be spectacular, the movement is east to west, across the bay from Kent Island. In both seasons the best conditions are on a day of light southwest winds three to four days after a strong cold front.

A narrow trail in the marsh between the observation tower and the East Beach leads past a small pond hidden in the reeds, which are reasonably dependable for American Tree Sparrows in winter. Another half-hidden pond lies beside the south side of the East Beach parking lot. Least Bitterns and Common Moorhens sometimes breed in these marshes. Years ago the reeds were less widespread and the flats they have replaced attracted almost every shorebird on the Maryland list. These days, most are fly-overs, though there is still a lagoon or two they like when water levels are right. (The bay shoreline is worth checking, too, of course.)

Three nearby spots outside the park offer prime woodland birding:

1. As you exit from the park heading west, turn north on Log Inn Road. Just 0.5 mile from the access road next to U.S. 50, the road makes a right-angle turn and bends left at 0.8 mile. Park at the bend and take the footpath on the north side of the road, which will lead you to pine stands popular with owls, and beyond into fine deciduous forest.

2. The 210-acre Corcoran Environmental Study Tract is accessible to any birder who asks permission at the park office. The entrance lies 0.7 mile down the road from U.S. 50, on the left. (Keep straight on Tydings Road when Log Inn Road turns right.) The tract is fenced in and the gate is locked, but you will be told at headquarters how to get in. Mature pines, tulip poplars, and abandoned fields provide varied habitat for owls and passerines.

3. For an hour or two of good birding, follow the access road that parallels U.S. 50 on the south, east from Whitehall Road. Take Colbert Road, the first right, and continue straight on for 0.1 mile on gravel beyond the point where the paving goes left. Park by a little log cabin and explore the nature trail across the road and

the fields to the east. This sanctuary, owned by the Chesapeake Bay Foundation, is both charming and birdy, especially in migration. Access is unrestricted.

St. Mary's and Calvert Counties 21

St. Mary's County, at the southern tip of Maryland, used to be among the most underbirded areas near Washington, but its attractions have been recognized increasingly in recent years.

Though Point Lookout lacks the ornithological drawing power of the great concentration points along the coast, such as Cape May and Kiptopeke, like them it lies at the tip of a rapidly narrowing peninsula. The funneling effect of the Potomac River and Chesapeake Bay pushes land birds in fall migration into the pockets of woodland and scrub in the park at the end of the road. On any sort of westerly wind chances are excellent for hawks in good numbers overhead, while telephone wires along the farm roads invite vagrant flycatchers.

From late summer, when herons, gulls, and terns from the nesting islands across the bay come to loaf on the sandbars, pilings, and jetties along both shores, all through the winter and spring, you are likely to have a wealth of fine birding, with everything but variety in pelagics, dabbling ducks, and shorebirds to keep you happy.

Numerous detours to points along one side of the peninsula or the other are worth making, and there are two strategies for birding the county. One is to start birding almost as soon as you cross the county line, getting to Point Lookout in time for lunch. The other is to drive straight to the point and spend the morning there. The latter procedure, widely preferred, is likely to be most rewarding in fall migration on a day when winds are out of the northwest. In that case songbirds will be pinned to the vegetation at the point and raptors will be rounding the tip and then heading across the Potomac. A morning of low cloud or drizzle in May is also a likely one for passerines to be in the shrubbery. Otherwise, there is merit in meandering down the county.

It is about a two-hour drive from the Capital Beltway, I-495 (mile 0.0), via Md. 5 and 235, the most direct route. In 14.9 miles Md. 5 leaves U.S. 301 in Waldorf, a left turn easy to miss. In another 18.9 miles it forks right from Md. 235. The two routes reunite at Ridge, 7 miles from Point Lookout: Md. 5 after 33 miles, Md. 235 after 31 miles.

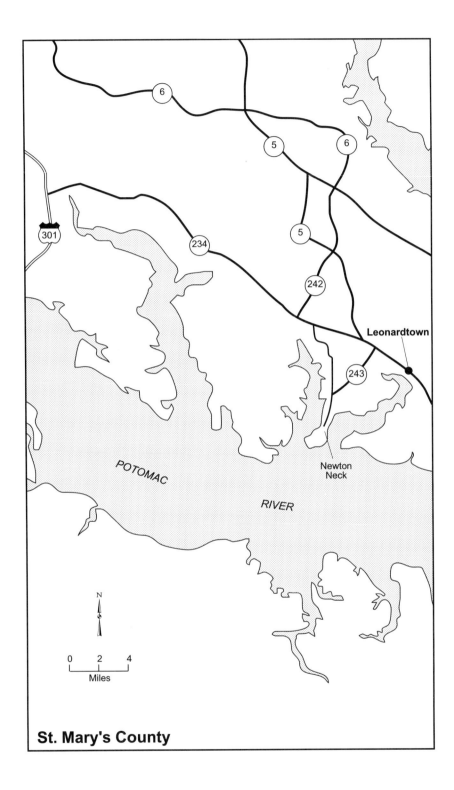

6

5

6

301

234

5

242

Leonardtown

243

Newton
Neck

POTOMAC

RIVER

N

0   2   4
Miles

St. Mary's County

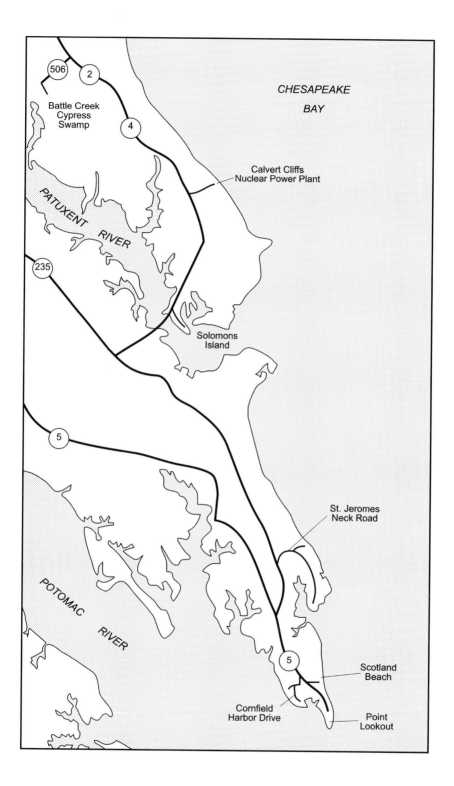

The first detour is before the divergence, 14.6 miles from U.S. 301. Turn left on Md. 6 and drive 2.5 miles to All Faith Church. Check the magnolias and oaks for migrant warblers and the field behind the church for wintering sparrows. Then continue 1.1 miles to Lock Swamp, where Wood Ducks and Hooded Warblers breed, kingfishers and bluebirds are conspicuous, and wintering sparrows lurk along the edges. (Do not leave the road.)

In another 1.5 miles, turn left on Trent Hall Road and drive 1.5 miles to the end of the paving, where you must turn around. The fields along this road are favored by Northern Bobwhites, Eastern Meadowlarks, and Horned Larks. When the trumpet vine is in bloom, Ruby-throated Hummingbirds are usually present. In the marsh Willow Flycatchers breed irregularly, and in the open water of Washington Creek, at the turnaround, bay ducks shelter in winter. Sharp-shinned and Cooper's hawks often dash by in fall.

Retrace your route and go on down Md. 6 to Partlett-Morgan Road, another left turn, and a short route good for wintering Savannah Sparrows and for Grasshopper Sparrows in March and April. You can see the other end of Washington Creek from the end of the road.

Back on Md. 6, marshy Persimmon Creek, a half-mile south, may have Blue-winged Teal when the water is up, Rusty Blackbirds and Acadian Flycatchers in season, and Hairy and Pileated woodpeckers at any time. The second road left, Delabrooke Road, goes to the Patuxent River bank. Park in the lot behind the restaurant. Swamp Sparrows live in the marsh across the road.

Md. 6 meets Md. 235 in 1.3 miles. Turn left to go straight to Point Lookout or continue across the highway for 7.2 miles, crossing Md. 5, and turn left on Md. 234. In 0.8 mile turn right on Bayside Road for **Newton Neck.** Along this road you may see Northern Harriers, Eastern Bluebirds, Blue Grosbeaks, Grasshopper Sparrows, and Eastern Meadowlarks. In 0.8 mile park carefully at the bend and walk down the road 0.2 mile to the creek. Barred Owls, Cedar Waxwings, Acadian Flycatchers, and Parula Warblers all nest in the vicinity. The bottomland woods a little farther on are worth a stop as well.

In another 3 miles turn right at the T-junction on Compton Road, which soon opens up to St. Mary's River on the left, open fields on the right, possibilities of Snow Buntings in winter, Vesper Sparrows in November and April, and Bald Eagles and Ospreys overhead. You will pass a church and a manor house that were built in 1640 and reach the end of the paving in 2.8 miles. Turn around there, but bird the area before you leave. Pine and Yellow-throated warblers breed here, as well as a dozen more common species.

Returning up Compton Road you will reach Md. 5 in 5.7 miles. Turn right and go directly down to **Point Lookout,** keeping an eye out for the resident Bald Eagles in the vicinity of the village of Ridge. Stop at the park administrative office and ask for two maps, one of the park, the other of St. Mary's County. If the office is closed and the campground is open, you can pick up the maps at the entrance gate, just ahead on the right.

Be warned that "No Parking" signs and traffic laws are strictly enforced in the park. "Do Not Enter" applies to vehicles; "No Trespassing" and "Area Closed" applies to you.

At the end of the road, park beside the rest rooms just above the tiny naval station. You can get good views, at either end of the fence around the latter building, of the sandbar and pilings beyond, favored by gulls, terns, and cormorants, both Great and Double-crested.

The river and bay, here as elsewhere in this account, may have from late fall to early spring hundreds of Horned Grebes, Tundra Swans, scaup of both species, Canvasbacks, Buffleheads, Common Goldeneyes, Oldsquaw, all the scoters, and Ruddy Ducks.

Just above the parking area, check the ponds and marsh in season for herons, long-legged shorebirds, Seaside Sparrows, and the only nesting Boat-tailed Grackles on the Western Shore.

You can easily spend most of a day in spring and fall birding all the woods or isolated trees and bushes for land-bird migrants. The park not only attracts a wide variety of warblers, but is also a magnet for thrushes, including Gray-cheeked, and *Empidonax* flycatchers, including Yellow-bellied. A walk north along the seawall to the east leads past a pine stand with resident Great Horned Owls and Pine Warblers to a brushy area that can be alive with sparrows. Above the picnic area to the west the woods are crisscrossed with perfect trails for birding, and the south side of the boat-launch parking lot should be checked as well, especially the cove and marsh in the southwest corner. The picnic area gates are open from May through October, and an entrance fee—$5 for a car, $1 for a pedestrian—is charged daily in summer, and on weekends before Memorial Day and after Labor Day.

Park next at the fishing pier and scan the bay for almost any kind of water bird in season. Gannets are most likely in early spring; even Wilson's Storm-Petrels are possible in summer. Then leave your car next to the seawall at the north end of the fishing area and walk with your scope up the beach to Tanner's Creek. Look over the jetty into the bay and inspect the cove, its shoreline, and its sandbars for herons, swans, ducks, terns, gulls, and perhaps a few shorebirds. In late summer sort through the Royal Terns for a rare Sandwich Tern.

Visit the museum if it is open, look for Brown-headed Nuthatches along the hiking trail or in the campground if it is fairly empty, and bird the north shore of Lake Conoy if the area is not overrun with people. If the campground is closed, you may park outside it, if you do not block the gate, and walk in to bird.

Back on Md. 5, go left 0.8 mile north of the park onto Cornfield Harbor Road, and left again onto Cornfield Harbor Drive, marked by cement obelisks at the entrance. Park near the bridge 0.8 mile ahead. In the appropriate season, King Rails and Seaside Sparrows have been seen in the marsh on either side, and Redheads have been observed with Canvasbacks in the creek to the east. Woodcock have been found in the woodland edge of Cornfield Harbor Road on the way back to Md. 5.

At the junction of Md. 5 and Md. 235 take Md. 235 to the right. After 2.7 miles, a right turn on St. Jerome's Neck Road will lead you eventually back to the bay again after winding for 4.6 miles through open country largely occupied by a nursery, likely territory for wire-perching birds and sparrows. Turn right at the T-junction just beyond the nursery, and at the end of the road park on the left in front of the "STOP–Eberly–No Trespassing" signs. Walk out to the bay on your left. At low tide the beach is broad and attractive to migrating shorebirds.

If you continue north on Md. 235, it will eventually rejoin Md. 5, but an alternate route home is Md. 4, 14 miles up from St. Jerome's Neck Road. Fork right on Md. 4 and in 3 miles take the high bridge over the mouth of the Patuxent River. The first right turn on the other side leads into the town of **Solomons,** where you can bird the river and also visit a marine museum and have a seafood dinner. To get to the boat launching ramp under the bridge, a good vantage point, take the first possible right, just beyond the visitor center. The river is also easy to study from the long parking lot on the road into town. Loons, grebes, and waterfowl—especially bay and sea ducks—may be abundant in winter and early spring.

If you still have daylight to squander, stop at the Calvert Cliffs Nuclear Power Plant, open daily 10 A.M. to 4 P.M., with fine views over the bay at the point at which the warm water from the plant is discharged into it. The area of upwelling can be a spectacular magnet for water birds in very cold weather.

Battle Creek Cypress Swamp Sanctuary, west of Md. 4 1.5 miles north of Port Republic, is especially recommended for its spring wildflowers and breeding warblers: Kentucky, Yellow-throated, Hooded, Worm-eating, Parula, and Prothonotary. There is a long boardwalk through the beautiful swamp, an upland trail, and a nature center. Turn west on Md. 506 and then left in 1.8 miles onto Grays Road. The entrance is 0.25 mile ahead, on the right. The sanctuary is open from April to September from 1 P.M. to 5 P.M. on Sunday and from 10 A.M. to 5 P.M. from Tuesday to Saturday. The rest of the year it closes at 4:30 P.M.

The familiar brown signs draw your attention to several other parks off Md. 4 on both sides of the road; take a day sometime to explore them on your own.

When Md. 2 forks right for Annapolis 27 miles north of Solomons, take it 4.5 miles to Md. 260, and turn right. In 4.4 miles you will reach **Chesapeake Beach,** a resort community on Chesapeake Bay. From November to April it is an easy place to see Tundra Swans, Canvasbacks, Greater Scaup, Oldsquaw, Buffleheads, Common Goldeneyes, and assorted gulls, with loons, Horned Grebes, and scoters passing by in migration. (Not all of these species are likely to be seen on any one visit.) In March and April an Eared Grebe or two is found almost every year, and dozens of Northern Gannets may be close enough to shore to enjoy with binoculars. Purple Sandpipers and a Red-necked Grebe have been recorded there but cannot be expected.

If you go south on Md. 261, the shore highway, for 0.3 mile, you can pull up at the water's edge in the parking area for the Rod 'n' Reel Restaurant. Scan the bay and the birds around the jetties on either side of the entrance to the little harbor. Then drive north 1.3 miles beyond the traffic light on Md. 261 to the Ann Arundel County line. Work your way south, keeping as close to the edge of the bay as possible. (For most of the way parking is not available on the east side of the road.)

To drive directly to Chesapeake Beach from Washington, go east from the beltway on Md. 4 about 6.5 miles beyond U.S. 301. Turn left on Md. 260, which ends in Chesapeake Beach at 15.2 miles. The four-lane divided highway all the way attests to the heavy traffic in summer, when birds are replaced by people. Come in the afternoon unless it is cloudy, and avoid days of easterly winds.

The Eastern Shore

At almost any time of the year the very best area to see the greatest diversity of birds in a single day's outing from Washington is a twenty-mile stretch of impoundments and marsh, fields, and woods near Dover, Delaware. There must be few avid birders in these parts whose life lists have not been substantially lengthened by happily remembered expeditions to **Bombay Hook National Wildlife Refuge** and Little Creek Wildlife Area.

From the tollbooths of the Chesapeake Bay Bridge (26 miles east of the Capital Beltway on U.S. 50), take U.S. 301 northeast about 34 miles and turn right on Md. 300, which runs east through farmland and small villages some 15 miles to Kenton, Delaware. At the traffic light in Kenton go right on Del. 42, which ends after 9 miles in Leipsic (pronounced *lip-sick*), crossing U.S. 13 en route.

Go left (north) in Leipsic on Del. 9 1.5 miles to the Bombay Hook entrance sign and turn right. In winter and early spring the fields on both sides of the road between the highway and the first farmhouse are sure to have Horned Larks in them and may have Lapland Longspurs. From mid-March to early May they are frequented by Lesser Golden-Plovers, which blend equally well with corn stubble and freshly plowed earth. Check any wet spots in the fields for other shorebirds. The entire entrance road is good for migrating and wintering sparrows, and the skies should be scanned for hawks, notably Rough-legs, Northern Harriers, and falcons.

The second road to the left is also rich in sparrows and runs straight into the best warbler spot in the refuge—at the (locked) back gate to Finis Pool—before it turns sharply left. In May and September it pays to bird these woods by the gate before going on to the refuge visitor center.

Bombay Hook is open from sunrise to sunset except when the refuge dikes are too soggy to drive on. At the visitor center you will find rest rooms, a staffed information desk, refuge maps, bird and mammal lists, and a sheet of recent sightings. The fields around the center are good for sparrows and Bobolinks in season and Ring-necked Pheasants all year.

The refuge roads run along and around three impoundments, Raymond and Shearness pools and Bear Swamp, all with observation towers, all with foot trails into nearby woods, all bordered on the east by tidal flats and marshes. The Boardwalk Trail near Raymond on the right side of the road takes you beyond the woods into interesting marsh habitat where you may find Sedge Wrens. Exten-

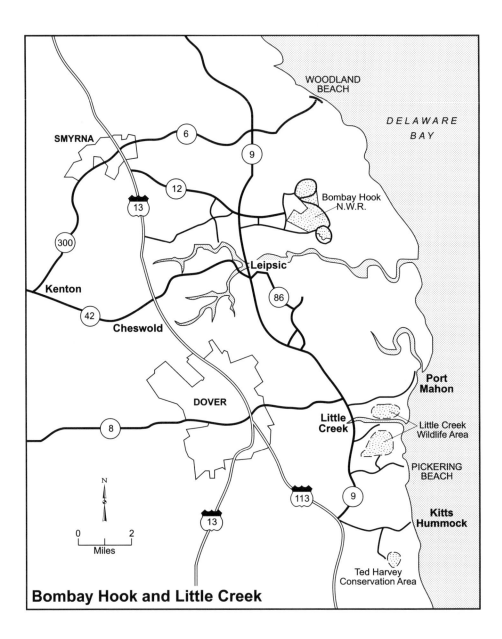

WOODLAND
BEACH

DELAWARE BAY

SMYRNA

6

9

12

13

300

Bombay Hook
N.W.R.

Kenton

Leipsic

42

86

Cheswold

Port
Mahon

DOVER

Little
Creek

Little Creek
Wildlife Area

8

PICKERING
BEACH

N

113

9

Kitts
Hummock

0   2
Miles

13

Ted Harvey
Conservation Area

Bombay Hook and Little Creek

sion roads lead to Finis Pool, essentially a huge swampy arm of Shearness that is good for Wood Ducks, and through cornfields to another locked gate.

The most famous resident species is the pair of Bald Eagles that usually nest on the west side of Shearness Pool but may be seen soaring anywhere. The best vantage point from which to see them is from the main road near the entrance gate. Look north, but do not enter the field. The eagles often perch at Parson's Point halfway to the far side of Shearness and are sometimes seen at low tide on the mud flats on the other side of the wildlife drive.

From September to April, with a lull in January and most of February, the numbers of waterfowl are spectacular, especially the tens of thousands of Canada Geese in late November and early December. (Look for Greater White-fronted Geese among them.) Snow Geese prefer the marshes, but the impoundments are packed with all the dabbling ducks, most of the diving ducks, and all three mergansers.

Shorebirds begin to appear in late March, are present until early June, and return again in July, continuing up to November. The tidal flats are always reliable, but in wet summers there may be no shorebird habitat in the impoundments until September. Raymond Pool is particularly noted for its Black-necked Stilts, its large, lingering flock of Avocets, and its attraction for Wilson's and Red-necked Phalaropes, all five "peeps," and an occasional Curlew Sandpiper.

Double-crested Cormorants, Common Moorhens, and a wide assortment of rails, herons, egrets, ibises, and terns show up in the warmer months. White Ibises are rare visitors most likely to appear from mid-July to (rarely) September. Woodland birding is excellent in migration in the woods north and west of Shearness and around Finis Pool. Blue Grosbeaks often nest near the Bear Swamp loop, as well as in the hedgerow along the entrance road. Be alert for the sight and sound of Barn, Eastern Screech-, Great Horned, and Barred owls, which are all resident. Look for Grasshopper Sparrows in the fields around the exit from Bear Swamp Loop and the beginning of the road to Finis Pool.

In winter only, when you leave the refuge (or before you visit it), go north on Del. 9 0.1 mile and take the first left fork toward Smyrna (Road 12). The second road on the left, Road 326 at 1.4 miles, takes you by a farmyard on the right, noted for Brewer's Blackbirds, and by fields that often hold American Pipits and sometimes Lapland Longspurs. Be sure to do all your birding from the road; trespassers are widely disliked. (The next left returns you to Road 12.)

Note: The Collections Manager at the Delaware Museum of Natural History in Wilmington has a special request. He urges that you bring to the museum any road-killed Brewer's Blackbird you find in the state and that you send him any identifiable photograph you take of one. After forty years of reports, all records of this species in Delaware are based on written descriptions only!

USEFUL FACTS

1. An entrance fee is required in all seasons. A pass that admits you to a national park or a current "duck stamp" serves the purpose. If you buy your stamp at the refuge, it will go to support the refuge itself.
2. At Bombay Hook the light for viewing the impoundments is best in the morning, though the tidal flats are backlit then. Afternoon birding is disappointing except on a cloudy day or when the tide is out or partially out and shorebirds are in migration.
3. Hunting is allowed in a large part of the refuge during much of the period from late October to late January, and the roads beyond Shearness Pool are frequently closed to visitors in those months, except on Sundays.
4. Birding is relatively dull in the refuge in June and early July.
5. Mosquitoes and biting flies may be absolutely terrible from June (or earlier) through September. Spray from head to toe before you enter the refuge.
6. If you would rather be deprived of money than sleep and still be at Bombay Hook for early morning birding, take note that Dover, 4 miles south of Del. 42 on U.S. 13, has several good motels in a wide price range and at least two all-night restaurants.
7. In Leipsic and Little Creek a policeman often waits for speeders on Del. 9, especially in the summer.

Little Creek Wildlife Area is the inevitable afternoon complement to a morning of birding at Bombay Hook National Wildlife Refuge. Only 8 miles down Del. 9 from Bombay Hook, it provides surprisingly different birds and habitats. Most vantage points ask for eastward viewing, so it rarely pays to work the area before noon except on a cloudy day.

Starting from the village of Leipsic, you can either zip straight down the highway, which carries too much traffic to dawdle on and inadequate shoulders to pull off on, or go part of the way via a very peaceful back road. Go east on Second Street, the extension of Del. 42, into the village to the stop sign on Main Street, and turn right. Then take the first left on Road 86, which starts out as Texas Lane, runs east, and bends south and finally west, rejoining Del. 9 just above the Octagonal Schoolhouse (a historical landmark), 2.3 miles north of the junction of Del. 8 and Del. 9 at the north edge of the village of Little Creek. This four-mile loop will take you past fields and pastures, often with rainpools, through a small swampy woods and across a tidal creek. Hawks, shorebirds, bluebirds, warblers, black-birds, Blue Grosbeaks, and sparrows are all possible in season; it is well worth exploring. Yellow-headed Blackbirds are found almost annually among the large flocks in autumn.

The wildlife area has three points of access. The northernmost is well-marked Port Mahon Road, the only road that runs east from Little Creek off Del. 9. About 0.9 mile east, opposite a large group of oil storage tanks on the left, a gravel road on the right leads to the northwest corner of the north impoundment (usually called the Port Mahon impoundment).

If cars are up on the dike (there is room for two), park in the area below. Otherwise you can park on the dike and bird from your car as you eat lunch, if so inclined. Walking the dike is easiest in winter and spring before the reeds grow high. From mid-July to early September, however, shorebirding is so good about a third of the way east along the dike that it is worth the hot and scratchy walk. The half-exposed mud flats to the south are among the best places to look for the Rufous-necked Stint that seems to show up nearly every year. The light is always bad except on an overcast day.

Like the south impoundment discussed later, this huge area, a mixture of open water, acres of marsh, and, sometimes, freshwater flats, is a paradise for dabbling ducks. (Look for Eurasian Wigeon when the American Wigeon are in.) Harriers, falcons, and Rough-legged Hawks hunt here; gulls rest and feed on the flats, with a rare Common Black-headed Gull or an occasional Little Gull among the Bonapartes. All the tern species on the Delaware list are likely to be here in summer. This, together with Bombay Hook, is the northernmost area in the East where Black-necked Stilts nest regularly and is a favorite shorebird concentration point in dry years. Sedge Wrens sometimes nest west of the north-south dike in the saltmeadow grass, while Marsh Wrens bubble reliably in the reeds.

If the tide is low enough to expose some flats, continue east on Port Mahon Road out to Delaware Bay. Check the swallows on the telephone wires in summer for an occasional Cliff Swallow. You will soon cross then run parallel to a creek that attracts herons and Clapper Rails. (Do not confuse their black chicks with Black Rails!) Scan the marshes to the north for egrets and ibis from spring to fall, Northern Harriers, Rough-legged Hawks, and Short-eared Owls (especially at dusk or on a dark day) in winter. The marshes are also prime habitat for Seaside Sparrows. (Look twice before you decide you have seen a Sharp-tailed Sparrow here; many of the ones reported after midsummer turn out to be juvenile Seasides.)

Immediately beyond the bend, the creek spills out into the bay. The flats here at low tide can be packed with every probable species of shorebird and gull (and, astonishingly, a Little Stint was photographed here in May 1979). The branch on the other side of the road lures rails out to feed until the traffic gets too heavy or the tide too high.

The poorly maintained bayside road is partly protected by a stone and metal seawall. Along the open shore you are likely to find shorebirds, especially peeps, Red Knots, and Ruddy Turnstones. In winter the fishing pier provides an ideal platform for viewing diving ducks and sea ducks and gulls out in the bay. Explore the shoreline for about a mile beyond the creek mouth; you can turn around at the end of the road. To the east a marshy point jutting south often provides a resting place for Royal and Common terns.

On your return to Del. 9 a left turn takes you after 1.5 miles to the main entrance to the Little Creek Wildlife Area, on the left. (The sign for the entrance is easy to miss at 50 miles an hour.) Follow the entrance road (frequented by wintering White-crowned Sparrows) past the headquarters buildings (there is no reason to stop) about 1.3 miles to a small parking lot. A boardwalk leads to an observation tower over the south impoundment, with much more water and much less marsh than in the north impoundment. Consequently, this is the most likely spot to see huge concentrations of waterfowl—most spectacular in March, when drifts of Snow Geese swirl along the far shore as the first Glossy Ibis coast in from the south, over crowds of Ruddies, mergansers, scaup, and a dozen other species of duck.

A closer look is available from the road beyond the parking lot, which is open to pedestrians despite the "No Entry" sign and leads to more ill-defined and overgrown dikes extending into both north and south impoundments. In drought summers there can be stupendous flats to the right, but you will be looking into the sun.

Back on Delaware 9 turn left; after 0.9 mile go left again on Pickering Beach Road. After 1.6 miles you will see a sign for Little Creek Wildlife Area on the left, with a locked gate across the entrance road. Pedestrians willing to walk a half-mile can reach the south edge of the south impoundment. You will be at the junction of two dikes, one running north and one running east out to the bay. Both provide access to far-off birds, but the latter also leads to a marsh area popular with waders (including both bitterns) and long-legged shorebirds when the water in the impoundments is too deep for them. You may find the Black-necked Stilts here when they seem to be nowhere else.

Pickering Beach, at the end of the paved road, is a good place to look over the gulls and ducks on the bay in winter; shorebirds like it at low tide, especially in May. Most of them will be 100 yards or so up the beach to the left.

The last stretch of Del. 9, 2 miles south to U.S. 113, is at its best when there are rainpools in the fields and the grass in the pastures is short. In September look for Lesser Golden-Plover, and from mid-July on, check for Upland Sandpipers in the field north of the junction of the two highways.

The Logan Lane Tract of the **Ted Harvey Conservation Area** lies 2.0 miles east of this intersection, off the road to Kitts Hummock. At the large sign, turn right (south) and drive a half-mile to a brick house next to a gate that may be locked. If it is, you will have to park your car and walk, but it is often possible to drive the 1.5 miles to a marsh and pond at the end of the road. At 0.5 mile, after another gate (usually open), follow the main road sharply left. In 0.2 mile you will pass a small picnic area with toilets. Continue to the marsh. Scout around for the best vantage point from here; it will depend on the water level and the size and location of the mud flats. You may walk a short distance on the track beyond the locked gate at the parking area.

This area achieved fame in 1988 when a White-winged Tern showed up here in late July. It, or another bird, was back in 1990, along with two Curlew Sandpipers. Yet even without rarities the area offers good studies of herons, shorebirds, and terns when conditions are right, i.e., when water levels are low enough (but not too low) for terns to rest and shorebirds to feed. Any sought-after water bird that disappears from Bombay Hook and Little Creek may turn up here. For that matter, it may have come here first!

If you're in the area near high tide, go on to Kitts Hummock, 1.1 miles beyond the turn-off for Ted Harvey. Shorebirding and views of the bay can be as good as at Pickering Beach, but the birds stay out on the far edge of the tidal flats, which can be extensive and inaccessible at low tide.

In the winter months, if you have covered Little Creek with an hour to spare before sunset, you may want to dash south to Broadkill Marsh, described at the end of Chapter 27. From the junction of Del. 9 and U.S. 113, drive south 11.8 miles, fork left on Del. 1, and continue 13 miles to Del. 16. Turn left and drive to the marsh.

From spring to fall, lovers of steamed crabs, the quintessential Delmarva feast, may want to head for Sambo's in Leipsic (closed on Sundays) or the Coral Reef in Little Creek for the ideal way to end the day.

23 Northern Delaware

Whether keeping state bird lists is one of the games you play or whether you simply like to expand your stock of splendid birding areas, northern Delaware offers a wealth of delights you shouldn't miss. It is worth noting that Wilmington is no farther from Washington than Bombay Hook, and traveling between the two destinations is but a leisurely, back-road drive of an hour or so, if you do not make the suggested stops.

The recommended route can be taken in either direction, but it is desirable to bird the Wilmington area in the morning if you can.

If you begin in Leipsic—the village just south of Bombay Hook—with a few detours you will be following Del. 9 north as far as Delaware City. This is a lovely back road, twisting through fields and marshes, passing patches of swamp and stands of pine, crossing tidal creeks, and going by handsome old farmhouses (interspersed with increasing examples of Nondescript Modern). It is at least as rewarding a road in fall and winter as in spring (owls in the pines, hawks over the fields, blackbirds feeding in corn stubble, and water-fowl in the creeks and marshes), and is always more interesting than bustling U.S. 13. If you take it between the Memorial Day and Labor Day weekends, however, you will be frustrated by the heavy traffic; it is being touted these days as an alternate route to the beach.

Measuring from the junction of Del. 9 and Del. 42 in Leipsic, after 7.5 miles you go through a half-mile of swampy woods that can be rich in warblers. Stop and bird the shoulders here if you have time. (At the end of the woods a left turn on Road 317 will take you to Boondocks, a good, plain seafood restaurant that is about the only place to eat en route.)

Continuing north, turn left on Road 456 in 7.6 miles and go 0.4 mile, as far as the wooden bridge over Beaver Branch. This stretch is likely to be good for Eastern Screech- and Great Horned owls and Red-tailed and Red-shouldered hawks, as well as migrant and nest-ing land birds.

Back on Del. 9, you will come to the Baxter Tract, a public hunting area 7.6 miles up the road, with a parking area on the west side of the road. This is another good place to look for owls; Long-eareds have been found here.

Out of season, Augustine Beach, 3 miles farther up Del. 9, is a good place to scan the Delaware River for waterfowl and look for sparrows in the edges. A little beyond lies the old village of Port Penn;

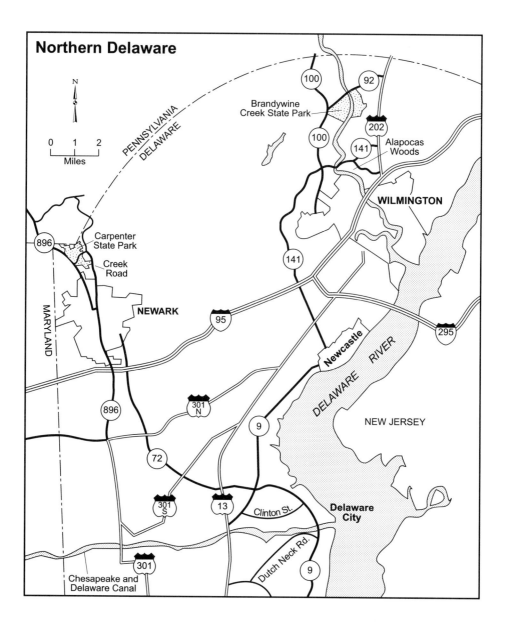

Northern Delaware

N

0   1   2
Miles

PENNSYLVANIA
DELAWARE

MARYLAND

100

92

Brandywine
Creek State Park

202

100

141

Alapocas
Woods

WILMINGTON

896

Carpenter
State Park

Creek
Road

141

NEWARK

95

Newcastle

DELAWARE RIVER

295

NEW JERSEY

896

301
N

9

72

301
S

13

Clinton St.

**Delaware
City**

301

Chesapeake and
Delaware Canal

Dutch Neck Rd.

9

2 miles north of it looms the high bridge over the Chesapeake and Delaware Canal. Just before the road rises to the bridge turn left at the sign for Dutch Neck Road. The mixed stand of pines and multiflora rose tangles along the east side of the bridge may shelter Saw-whet Owls in winter. You will soon come to the canal bank, where a right turn will lead to the south jetty of the canal entrance, an excellent spot for loons, grebes, and gulls in the colder months. On

your right, a long, brushy, man-made hill is excellent for sparrows in fall and winter and has been known to shelter the occasional Lincoln's or Clay-colored Sparrow as well as more common species.

A left turn along the canal takes you past the Thousand Acre Marsh, a vast freshwater pond to the south, which is a superb shorebird area in a dry summer (Ruffs and Marbled Godwits have been found here, as well as the more routine migrants) and a fine waterfowl lake in winter. From the T-junction at the canal a drive of 1.1 miles west on the paved road will take you past a marshy pond with water lilies and spatterdock, which is especially promising for Common Moorhens and Least Bitterns. The whole area is a loafing ground for gulls and terns in late summer.

Retracing your route, go over the bridge and into Delaware City. Out in the river is Pea Patch Island, home of a huge mixed heronry in summer. This wonderful spectacle is normally inaccessible, but Brandywine Creek State Park runs a trip to it two or three times a year in late May or early June. The schedule is announced sometime in March. Call the park at (302) 655-5740 to find out the dates. Only paid reservations for individual participants are accepted (no groups).

Dragon Run Park, 0.2 mile north of the traffic light in Delaware City, on the left, gives you a good view of Dragon Run Creek. Here you can enjoy the number and variety of herons, egrets, and ibis that come over from Pea Patch Island to use the marsh all summer, and the ducks that spend the winter here. Then retrace your route and turn right at the light on Clinton Street; 0.9 mile along this road, park and look out over Dragon Run Marsh on your right.

If you cross the road and climb the embankment on the left by means of a path just ahead, you will look down into a large impoundment popular with dabbling ducks and coots.

Continue another 1.8 miles, where, a little beyond the school, you turn right on unmarked Road 378. In 0.5 mile you will cross Dragon Creek. If you don't have to share it with fishermen you have a good chance to see King and Virginia rails and at least hear Soras, as well as other marsh birds and transient Bobolinks.

After 0.7 mile turn left on Del. 72, passing the Getty refinery, and turn right after 1.6 miles on U.S. 13. In the course of the next 8 miles you will find plenty of motels, but they fill up early and it is wise to have a reservation. It is the most convenient stretch to stop for the night, for the morrow's early birding, in one of two beautiful areas, Brandywine Creek State Park and White Clay Creek. (I should add immediately that the latter is the favorite land-birding spot of every Delaware birder I know, but that is partly because it is accessible at all hours and is always free. Try them both.)

In Wilmington your principal target will be **Brandywine Creek State Park,** which is easily reached by taking Del. 141 north (toward

Newport), a turn from the right lane off U.S. 13, 8.3 miles north of U.S. 72. Stay on Del. 141 for 7.7 miles. Go left at Del. 100 for 2.1 miles (until it makes a sharp left turn), and there turn right at the conspicuous park sign on Adams Dam Road. After 0.3 mile turn left on the park entrance road and drive to the nature center.

This wonderful park consists of 433 acres of rolling hills sloping down to Brandywine Creek. There are large stands of ancient tulip trees broken by overgrown fields, with maples, box elder, and tangles of honeysuckle along the creek, and an always birdy marsh in the southeast corner. The highest ground is kept mowed and a park road leads up to the well-marked Hawk Watch, often staffed by Delaware birders in the fall.

It and White Clay Creek are the two best places in Delaware for upland species, both migrant and resident, and the diversity of flycatchers, warblers, and sparrows is astonishing in spring and fall.

There is a steep fee for nonresidents from Memorial Day weekend to Labor Day, but entrance is free the rest of the year. The other restriction is that the park is not open until 8 A.M., but there is a convenient and productive area nearby to while away the earliest birding hours.

Continue on Del. 141 instead of turning left on Del. 100. Go 1.2 miles beyond the crossroad, turn right at a traffic light onto Alapocas Drive, go through the next intersection, and enter **Alapocas Woods Park** (unmarked from this direction). Bear left, and in the course of the next 0.9 mile, you will be driving along the edge of Alapocas Woods, a county park with two or three entrances leading to parking spots and trails. You can walk along several stream valleys that often provide shelter for migrant thrushes in the brushy understory; the deciduous woods overhead are noted for their warblers and vireos in spring migration.

Take note that Del. 141 is easily accessible from Interstate 95, whether you are heading toward or away from Washington, and that the intersection of Del. 141 and Del. 100 is only 6.2 miles north of the interstate. You can also reach it by taking Del. 202 north from I-95 and turning left on Del. 141 and left on Alapocas Drive.

The other prime site in Delaware for migrating and breeding woodland passerines is **White Clay Creek,** on the north edge of Newark (pronounced *new ark*), a college town 9 miles southwest of Del. 141 on I-95. Newark has several motels and numerous places to eat, so you may want to stay here if this is your choice for birding at dawn. It is also a highly recommended site for late afternoon and evening birding.

This stream valley is about an hour and a half north of the Capital Beltway (I-95/495) in non–rush hour traffic. If you are coming from the south you can avoid a highway toll by exiting at the Newark/Elkton exit on Md. 279 and heading toward Newark. When you cross the state line the highway becomes Del. 2. In Newark the traffic pattern forces you to turn right on Delaware Avenue. Get in the left lane and turn left on Del. 896 (South College Avenue) into the *right* lane. At the T-junction with East Main Street, which is one-way west, turn left, and then turn right immediately on North College Avenue. This road leads into White Clay Creek at the north edge of town. (If you are coming down the interstate from the northeast, exit toward Newark on Del. 896. You will reach the T-junction with East Main Street in 2.9 miles.)

The extension of North College Avenue is known locally as Creek Road but is sometimes labeled Tweeds Mill Road on official maps. Its traffic is typically confined to birders, joggers, bikers, and (mainly in April) fishermen. There are a few places where you can pull off on the east shoulder along the creek, but perhaps the best place to leave your car is 3.0 miles north of East Main Street, where the creek road goes straight and becomes unpaved and the paving follows an unmarked road left. (At the other end, this road is signed Wedgewood Road, but some maps call it Appleton Road.)

All along the creek there is abundant habitat for thrushes, vireos, and warblers on the move. There are resident Barred, Eastern Screech-, and Great Horned owls, and Pileated Woodpeckers. The west side of the road in the quarter-mile south of the junction with Wedgewood Road is a wet, thickety area especially attractive to migrating Lincoln's Sparrows. A pair of Broad-winged Hawks nests nearby. A trail goes up the slope in the southwest quadrant of the junction to breeding habitat for Blue-winged and Prairie warblers and eventually to a field of tall grass where Grasshopper Sparrows nest. If you walk up Wedgewood Road a bit, just beyond a bridge you will come to a service road on the right that leads into Carpenter Park (which occupies much of the area). Walk down it to look for Worm-eating Warblers.

When you are ready to move on, drive north from the junction. The pumping station on the far side of the creek marks the beginning of Cerulean Warbler territory. At 0.7 mile from the junction a rewarding path beside a little stream to the left leads to Hooded and Kentucky warblers and loops back to the road.

In another 0.1 mile you will cross Hopkins Road, then a bridge that shelters a pair of Eastern Phoebes, and come to a parking area on the left where you must leave your car if you wish to continue farther. You may hear Ring-necked Pheasants nearby. North of this point the river banks become suitable for waterthrushes and Spotted Sandpipers. (Louisiana Waterthrushes and Belted Kingfishers breed here, as do Wood Ducks.)

Continue on foot past an open field to Chambers Rock Road and detour to the right along the road at least as far as the creek, a stretch that is very good for migrants. You may continue up the creek along a trail on the east side (highly recommended) or come back and keep on going up the road you have been on, as far as you like. (This trail is also very productive.) Ignore the well-preserved but out-of-date "No Trespassing" signs. This beautiful beech-hickory-oak forest will lead you on and on. Eventually you will walk into Pennsylvania.

If you go back to Wedgewood Road, it will lead you to Del. 896 north of town. Turn left to go back to Newark. If you turn right, you will come to the entrance to Carpenter Park (open the same hours and with the same fee schedule as Brandywine Creek) in 0.4 mile. Follow the road to a parking lot and rest rooms. Off the entrance road to the left a service road leads to a high open slope with a picnic table or two, where hawk watching can be as good as at Brandywine Creek.

Kent County, Maryland 24

Among the richest birding areas in the colder half of the year, Kent County is no farther from Washington, D.C., than popular refuges like Bombay Hook and Blackwater, but is much less frequently visited. Waterfowl, raptors, blackbirds, and sparrows and their kin can occur in spectacular abundance, and the rolling, open country of the upper Eastern Shore, dotted with handsome old houses built by three centuries of farmers, adds an aesthetic dividend to the exhilaration of seeing birds.

The starting point is Chestertown, 64 miles from the Capital Beltway, I-95/495, via U.S. 50, U.S. 301, and Md. 213. Cross the bridge over the Chester River, and go past the Washington College campus on your left. Start measuring your mileage as you turn left onto Md. 20. (Mileage figures assume you will take all recommended side roads.)

At 1.8 miles a sharp left turn onto Brice's Mill Road will take you down an ancient sunken farm road with sparrowy hedgerows. It ends at a T-junction at 3.7 miles; turn right, and at 5.1 miles continue across a tricky intersection that jogs slightly left. Look for feeding Canada Geese and Tundra Swans in the fields beyond and Red-tailed Hawks and Bald Eagles in the sky.

Park by St. Paul's Church at 7.3 miles, do a little land birding around the spillway you have just passed, and check the pond behind the cemetery for ducks. In the spring Warbling Vireos nest in the churchyard.

Go straight on (do not turn right around the church) and at 7.6 miles turn left onto the road to the main pond of **Remington Farms,** well

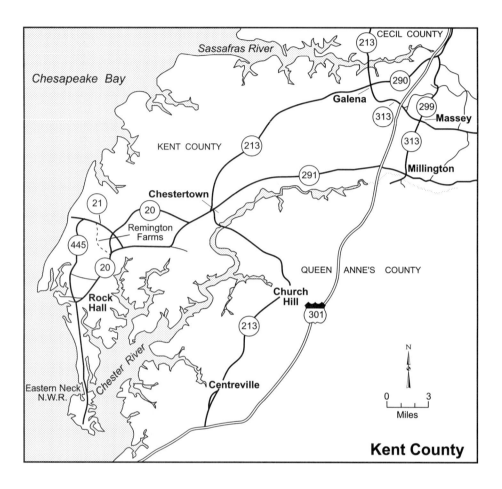

Kent County

signed and immediately visible. Remington Farms is a private water-fowl management area, but this part is always open to the public and can be packed with Canada Geese and dabbling ducks. A Greater White-fronted Goose may often be present, and such rarities as Garganey and Barnacle Geese, of unknown provenance, have been found in recent years. The light is good only in the morning. You are required to stay in your car, from which viewing is excellent and photography easy.

Make the mandatory U-turn, go back to the road, and turn left. The high hedgerows are full of White-crowned Sparrows and other birds, and the area abounds in deer at twilight.

At 9.4 miles the road ends at Md. 20. Across the road, a little to the left, is the entrance to the Remington Farms tour, open except from October 1 to February 1. The tour deserves at least an hour of your time, as it is delightful for buteos, ducks, and sparrows in winter and for nesting Ospreys, Blue-winged Teal, Wood Ducks, and migrating warblers in spring. You may be lucky enough to see a King Rail, but they are as elusive here as elsewhere.

The tour ends on Md. 21, on which a right turn takes you in 0.8 mile to Md. 20, where another right turn takes you back to the starting point of the tour in 1.1 miles.

If the tour road is closed, you will have turned left on Md. 20 (at 9.4 miles). At 13.2 miles turn left at the light in the village of Rock Hall on Md. 445 (Main Street), and at 14.4 miles take a detour right on Allen's Lane. Beyond the houses the woods on the right have a resident pair of spooky Great Horned Owls and are also good for Eastern Screech-Owls, woodpeckers, Hermit Thrushes, and Pine Warblers.

The lane ends at 15.1 miles at the bay, where you can scan for loons, Tundra Swans, both scaup, Canvasbacks, Oldsquaw, and all the scoters. Back on Md. 445 continue south, looking for bluebirds, and turn left at 16.0 miles on Grays Inn Road, a short spur to a marsh and public landing especially good for grebes and sparrows.

The highway now runs between narrow strips of wood and hedge-row that are always birdy, especially in migration, when they offer the best cover on the shoreline. At 20.4 miles you come to the bridge over Eastern Neck Narrows, the boundary of **Eastern Neck National Wildlife Refuge.** Park on the far side and linger to study the resting gulls (Lesser Black-backs and, much more rarely, a Glaucous or Iceland Gull are sometimes seen among the Ring-bills) and the shorebirds on the tidal flats, Tundra Swans, Canada Geese, Ruddy Ducks, Canvasbacks, wigeon (both American and, rarely, Eurasian), and both scaup. You may squeak up a Marsh Wren.

Drive on to Tubby's Cove on the right at 21.1 miles, where a boardwalk leads out to an observation tower. This walk provides the best views of the resident Mute Swans, all three mergansers, and Virginia and occasionally Black rails. It is a wonderful place to watch migrating raptors and flocks of passerines. You may also want to take Boxes Point Trail—walk back 0.1 mile on the road and you will see it on your right.

At 21.6 miles turn left on Bogle's Wharf Road, which leads down to a county-maintained public landing. Here you have fine afternoon views over the Chester River, especially attractive to Common and Red-throated loons, Bufflehead, Common Goldeneye, and scoters. Numbers and variety can be outstanding in spring. Check the waters in the cove as well. The little marsh on both sides of the road is good for Virginia Rails. Yellow-throated Warblers nest in the adjacent woods.

The Duck Inn Trail, on the left just after you turn onto Bogle's Wharf Road, goes through a good area in which to see Red-headed Woodpeckers.

At all times in the refuge remember to look for raptors, both in flight and perched. Both species of vulture; Red-tailed, Red-shouldered, and Sharp-shinned hawks; Bald Eagles; and Northern Harriers are reliable winter residents. Golden Eagles and Cooper's and Rough-legged hawks are possible; Ospreys are present from spring to fall.

Ingleside Recreation Area at 23.3 miles is closed from October to April except to those who get permission and a key from refuge headquarters, which is closed on weekends. It is worth a visit for its northward views over the bay, especially for Tundra Swans, scoters, and Oldsquaw.

The next road to the right, at 23.5 miles, leads up through sparrowy fields to refuge headquarters, a working office rather than a visitor center. If you have the time, ask about access to the trails to Cedar Point and Shipyard Creek, interesting open-country walks in the southeast part of the refuge. At least drive to the end of the public road, a high point from which raptor watching can be excellent. Retrace your route to Rock Hall and start measuring mileage again from the intersection of Md. 445 and Md. 20 (mile 0.0). A right turn on Md. 20 will take you straight back to Chestertown, passing the entrance to the Remington Farms tour road.

Alternatively, go straight north on Md. 445 through fields attractive to winter flocks of blackbirds, where the sharp-eyed may pick up an occasional Brewer's or Yellow-headed blackbird among the Red-wings and cowbirds.

At 1.0 mile there are woods with responsive nighttime Eastern Screech-Owls. Another good owling road runs east at 2.0 miles, but the next daylight stop is the marsh that borders the left side of the road from miles 2.6 to 3.0. Woodcock display here in the spring, and rails and egrets may be around in the warmer months.

Check the snags off to the right at 3.5 miles for Bald Eagles and take note that the woods at 4.2 miles shelter both Chuck-will's-widows and Whip-poor-wills in spring and summer.

You can drive down to the bay shore at 5.5 miles by turning left on Md. 21 or get back to Md. 20 and thence to Chestertown by turning right. The first mile east on Md. 21 can be wonderful for migrant warblers in season.

The opposite end of Kent County also deserves a visit in both summer and winter. It is easily combined with a visit to Bombay Hook National Wildlife Refuge. Take U.S. 301 beyond Md. 300, the usual turn-off for Bombay Hook, about 7 miles to Md. 291 and turn right. In 1.5 miles you will reach Millington. Turn left on Md. 313 and continue north on Md. 299 when Md. 313 goes left in Massey. Md. 299 ends on

Md. 213, 7.9 miles north of Millington, and a left turn will take you back to U.S. 301 in 0.1 mile.

This drive is best taken in winter, when Tundra Swans, Canada Geese, and Snow Geese (over from Bombay Hook) may be feeding in the fields and Rough-legged Hawks and Northern Harriers are hunting over them. On dark afternoons or late in the day Short-eared Owls have been seen regularly along the stretch between Quinn Road and Alexander Road, starting 1.6 miles north of Massey. Every bare field should be scanned for Horned Larks, American Pipits (in migration), and sometimes Lapland Longspurs and Snow Buntings. They may come to the edges of the road when the earth is largely covered with snow and the roads have been freshly plowed. Look for great swarms of blackbirds, among which you can try to find a Yellow-headed Blackbird if you are very lucky and have the necessary patience.

In early summer, cruise the side roads off Md. 313 and 299 in search of breeding Dickcissels and Vesper Sparrows. Recent reports of Dickcissels are from Alexander Road to the west and Hurlock's Corner Road to the east, between the ponds in the first half-mile.

Millington Wildlife Management Area lies along the state line, east of Massey and northeast of Millington. It is a good area for migrant warblers and other songbirds. Take any road east of Md. 313 and north into the refuge.

Note: The landowners of Kent County are not hospitable to trespassers. Ask for permission whenever you feel you must invade private property.

Blackwater and Hooper Island 25

Among the classic winter field trips for the Washington birder is an all-day expedition to Blackwater National Wildlife Refuge and Hooper Island in Maryland's Dorchester County. The best period of all runs from mid-November to Christmas, though birding can be good much longer in a mild winter and is interesting most of the year.

Your starting point is the junction of westbound Md. 16 and U.S. 50 just south of Cambridge, where there is a conspicuous sign for the refuge. Three roads—Maple Dam Road, Egypt Road, and Md. 335—run south from Md. 16 to the refuge. Egypt Road, 2.7 miles west of U.S. 50 and the first left turn beyond Cambridge High School, is the traditional birder's route; it has the virtues of being mostly straight and untraveled and lined by fields attractive to swans and geese, Horned

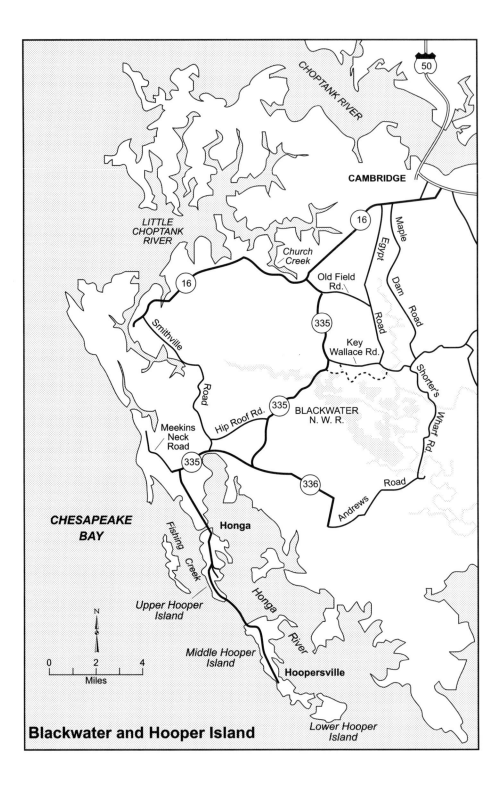

Blackwater and Hooper Island

Larks and American Pipits, blackbird flocks and sparrows, and, in a wet spring, a variety of shorebirds. Look for Eastern Bluebirds and American Kestrels on the wires and raptors in the skies.

In spring and early summer stop in the swampy woods 3.5 miles down Egypt Road for warblers (Prothonotaries nest there) and take a detour to the west down Old Field Road 0.8 mile farther along (it continues to Md. 335) for more mixed-woodland habitat that sometimes produces a Red-headed Woodpecker.

The last half-mile or so of Egypt Road is good for predawn owling (Eastern Screech- and Great Horned owls may respond to a tape).

Key Wallace Drive runs more or less along the north boundary of the refuge from Maple Dam Road to Md. 335. In both directions it is a fine road from which to see Eastern Bluebirds. In the winter, look for Eastern Phoebes and Pine and Palm warblers wherever you see bluebirds along the woodland edge. If you have arrived after 7:30 A.M. (9 A.M. on Sunday) for your first visit, go right 1.3 miles to the Wildlife Interpretive Center, pick up a refuge leaflet and a bird list, and look at the sighting list and the excellent tabletop map.

Thousands of geese may be just beyond the building. Snow Geese, both blue and white, will be easy to spot among the Canadas, but a rare White-fronted Goose and an even rarer Ross's Goose are worth searching for.

Next, head east 1.6 miles to the start of the Wildlife Drive, a 5-mile route. The first pond, on the right, may have good shorebird habitat in late summer and early fall and is outstanding for dabbling ducks and geese later in the year. (Since it is to the west you will want to have the morning sun at your back.)

A left turn beyond the first pond takes you by a picnic area and a nature trail, out to the vast marsh that borders the Blackwater River. Look for Bald Eagles soaring or in the marsh, Common Mergansers in winter, and a variety of terns—Forster's, Caspian, and sometimes Royal—in summer. The pine woods around the picnic area often shelter Pine Warblers and Brown-headed Nuthatches, but you should not expect to see a Red-cockaded Woodpecker in spite of its optimistic inclusion on the refuge bird list. (In recent years they were reported only in 1976, in a closed part of the refuge several miles away.)

Return to the drive. Beyond a woodland nature trail you will be in sight of water most of the way, much of it covered with geese, with more of them filling the skies. Among the dabbling ducks, Mallards and Blacks, Green-winged Teal and Shovelers, Pintail and Wigeon and Gadwall can all be sorted out; the attentive hunter of rarities scans diligently for Eurasian Wigeon. Diving ducks are less common, but may be seen in the ponds at the beginning of the drive.

The abiding excitement of visiting Blackwater is not the seasonal certainty of waterfowl, however, but the probability of seeing some of the resident Bald Eagles. They are easiest to see when they are perched on the snags along the Wildlife Drive, but they are as likely to be sitting out in the marsh or soaring out in the immense sky. It is important to check every birdlike lump in the trees at the edge of the marsh or out in the peaty flats.

The Blackwater area is superb for raptors generally. Northern Harriers and Red-tailed Hawks winter there, and Sharp-shinned and Cooper's Hawks are often visible from the Wildlife Drive; Ospreys nest there, and Golden Eagles show up every year, though it is easy to mistake immature Bald Eagles for them. Turkey Vultures are abundant, and Black Vultures not uncommon.

After completing the Wildlife Drive, loop back to the east again along Key Wallace Drive, past the refuge headquarters building. The marshy flats on the left (at low tide) are the best place to see wintering Common Snipe; they also attract other shorebirds, gulls, terns (in season), ducks, and herons, and the surrounding reeds shelter Marsh Wrens.

Key Wallace Drive ends at a T-junction; to the left the road is known as Maple Dam Road, but to the right it becomes Shorter's Wharf Road. Take your mileage at this point (mile 0.0). Turn right and bird the logged-over woods on the left for bluebirds, woodpeckers, and, in summer, Blue Grosbeaks.

The road soon leaves woods and farmland and runs through a magnificent stretch of marsh and pine that not only provides another chance to look for all the refuge specialties but also includes fine rail and shorebird habitat, especially in spring if water levels are not too high. In winter look for Northern Harriers, Rough-legged Hawks, and Common Mergansers. Farther south the habitat becomes suitable for Prairie Warblers and Field and Swamp sparrows. Having mysteriously become Andrews Road, the highway ends, at 10.1 miles, at Md. 336.

Turn right and head west and, at 14.9 miles, south on Md. 335. This road leads to Hooper Island, which is really a chain of islands, covered by small watermen's communities, interspersed with coves and harbors and extensive views of Chesapeake Bay. Explore every side road you see (they are few and short) and scan the waters wherever you can. In winter, loons, grebes, swans (including the resident Mute Swans), diving ducks, Oldsquaw, scoters, mergansers, gulls, and eagles may be seen anywhere. Terns and oystercatchers and a variety of herons nest on islands west of the villages of Honga and Fishing Creek, the nearest of which can be scanned from the road on a clear day. From the northernmost bridge to the end of the road is less than 9 miles.

Heading north again, 2.7 miles above the Hooper Island bridge, a left turn onto Meekin's Neck Road by a red-painted shed will take you all the way to the bay if the last quarter-mile is not too sodden. Most of the way is firm and dry. It is a good road for land birding, for hawk watching in migration (notably over the great marsh on the south side), and for observing duck movements up and down the bay (keep right at the Twin Willows sign).

Back on Md. 335, go left on Smithville Road 1.5 miles farther north and, a half-mile later, right on Hip Roof Road, a pretty stretch of marshes, woods, and fields especially good for herons and raptors. At the end, turn left onto Md. 335 again and follow it all the way back to Md. 16 in the village of Church Creek. Though you may encounter a lot of traffic on this route, you can usually find a shoulder to pull off on beside the marshes. Farther north, next to the fields, which may be seething with flocks of blackbirds, a patient searcher may some- times find a Yellow-headed or a Brewer's. If it is too dark for black- birds, linger along the marshes instead and listen for rails and Marsh Wrens.

Beautiful as this country is, you will find it marred by more "No Trespassing" signs than anywhere else in the world. (My favorite adds "Survivors Will Be Prosecuted.") Obey them fervently.

When goose hunting starts after Thanksgiving, avoiding the inces- sant sound of gunfire on Saturdays will enhance your pleasure con- siderably. Any other day of the week is better for a visit.

Elliott Island and a Detour from Easton 26

If you first visit Elliott Island when the sun is high on a summer day, its compelling attraction for birders will mystify you. Elliott Island, like all the great Delmarva marshes, is beautiful and full of birds, but in spring and summer it requires more of its devotees than most non- birders could imagine anyone submitting to.

The best hours to be there are from about two hours before sunset to two hours after sunrise, including at least an hour or two in full darkness. The most sought-after bird in the marsh, the Black Rail, is least difficult to see and hear (though never a sure bet) between 10 P.M. and 2 A.M. in May and June.

Not only must you turn your schedule upside down, but you will also need special equipment, the most important item being super-strong insect repellent, best applied before you dress and again just before you reach the marsh. You will in addition need a tape recorder, a tape of rail calls (preferably with long sequences of each species), possibly a tape of the calls of other birds you want to see, and a good flashlight.

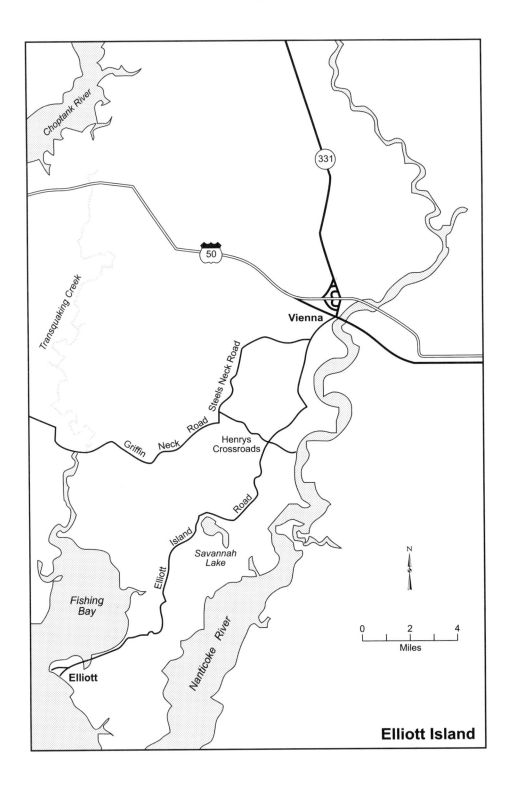

Choptank River

331

Transquaking Creek

50

Vienna

Steels Neck Road

Griffin Neck Road

Henrys Crossroads

Road

Elliott Island

Savannah Lake

N

Fishing Bay

Nanticoke River

Elliott

| 0 | | 2 | | 4 |

Miles

Elliott Island

Why go through all this effort? To see and hear birds you may encounter in no other way and to savor the beauty of a unique landscape.

If you cannot be persuaded to sample Elliott Island's summer pleasures, wait until November or later, when there are no mosquitoes, and Rough-legged Hawks, Northern Harriers, Bald Eagles, and waterfowl can all be seen in the middle of the day. Your trip can easily be combined with a visit to Blackwater National Wildlife Refuge.

The road to Elliott Island begins in Vienna, Maryland, off U.S. 50, halfway between Cambridge and Salisbury. Take the exit onto Md. 331 for Vienna. In 0.5 mile, after crossing old U.S. 50, turn left from Gay Street (the extension of Md. 331) on Race Street, and then take the next right on Market Street at the bank. From this point (mile 0.0) to the end of the road is a 20-mile run and the only decisions you need to make concern the stopping points.

The first 8 miles run through fields and deciduous woods. Stop, look, and listen frequently for Whip-poor-wills by night and Summer Tanagers and Blue Grosbeaks by day, and notice the intersection at Henry's Crossroads at 5.6 miles. (It is actually the road on the right that is called Henry's Cross Road, while the one to the left is Lewis Wharf Road.)

From this point on be punctilious about staying on the public road, and avoid parking beside any occupied house. Residents here find birder behavior threatening or irritating or both, and courtesy, inconspicuousness, and scrupulously lawful behavior are essential. After the Dietrich property at 7.8 miles you will be in loblolly pines, through which Brown-headed Nuthatches wander. Chuck-will's-widows call after dusk; they rest on the road and their red eyes may be picked up in the headlights. A tape of their call may bring one in to fly around the car. This stretch is a good one to hear Woodcock in April and May, near selectively logged areas, especially at the beginning of the marsh at 9.1 miles.

The waters of Savannah Lake on the left are normally birdless in summer (though herons may line the shore), but waterfowl are sometimes abundant there in the colder months.

Beyond the lake you are in rail country. If you do not have a tape, the slam of a car door may start them calling after nightfall, but only a recording of the voice is likely to pull nearby birds out onto the road.

Seaside Sparrows and Marsh Wrens sing far into the evening. If you arrive in daylight, look for the Marsh Wrens among the tall reeds along the tidal creeks that wind through the marsh, sometimes close to the road. Sharp-tailed Sparrows, though as common as Seasides, are more difficult to find. Look for them in late afternoon, when they are most active, in the expanses of short marsh-grass; Seasides prefer wetter areas with slightly taller grass.

The only nesting shorebird in the marsh is the Willet, but in migration the shallower pools along the road attract small numbers of other sandpipers—mostly yellowlegs, dowitchers, and peeps.

In the deeper ponds and creeks dabbling ducks, Common Moorhens, and Pied-billed Grebes are sometimes in sight, and herons and egrets fish along the edges. American Bitterns are scarce in summer but can be easy to find in October and November.

At dusk or dawn park out in the middle of the marsh, watch the changing light, look for the flights of egrets or night herons flying to or from their roosts, and listen for the booming of bitterns, the cuckooing of grebes, and the cacophony of the night birds and frogs chorusing against the chatter of the daylight singers.

The house on the left at 11.9 miles marks the start of one of the best stretches for Black Rails; the bridge at 13.6 miles marks the end of it, though they may be heard well to the north and south of those points on a calm night in a good year. Try to lure in the one that sounds closest to the road. In order to avoid damaging their habitat, do not leave the road to tramp around in the marsh.

Beyond this point the marsh is broken at intervals by islands of pines, home to House Wrens, Pine Warblers, Catbirds, Brown Thrashers, and Rufous-sided Towhees. At 15.4 miles a short road to the left runs through woods with a resident Great Horned Owl, where Yellow Warblers, White-eyed Vireos, and Common Yellowthroats sing along the edges.

At 16.6 miles the road runs close to the open water on the right. In daylight in the warmer months, terns sometimes hunt close to shore and in winter diving ducks can be observed at close range. Nest boxes for Barn Owls have been put up in the marsh here, and their inhabitants can be seen in late evening and before dawn.

You leave the marsh, cross over a bridge, and enter the outskirts of the village at 18.4 miles. Though it lacks architectural interest, it is attractively open and peaceful, and its residents show none of the hostility so manifest north of the marsh. Chuck-will's-widows fly back and forth across the road here at dusk, and a Barn Owl hangs out at the water tower.

At 19.0 miles, a road to the left takes you down to the harbor, whereas the main road ends close to the water at 19.7 miles in an interestingly swampy patch of woods. Both roads are worth taking for the opportunities they offer of scanning Chesapeake Bay, again much more rewarding in winter than in summer.

Elliott Island is not a place to bird in a hurry. Go prepared, and enjoy it. If you have arranged your trip to combine it with a visit to Blackwater, the linking road is the one to the left (as you head back to Vienna) at Henry's Crossroads.

Take every possible left turn (there are three more) except the one into the boat launching facility beyond the humpbacked bridge over Transquaking River. At high tide the road east of the river may be flooded, but rarely deep enough to prevent a slow transit. Stop on top of the bridge for a splendid view. In the marshes around the bridge you may find Rough-legged Hawks and Short-eared Owls in winter, rails in summer, and Sedge Wrens in late fall.

In 15.1 miles you will find yourself at the beginning of Key Wallace Drive, where a right turn will take you in 1.0 mile to the start of the refuge wildlife drive. En route you pass through fields, woods, and marsh, all good for birding from the road, especially for soaring Bald Eagles.

If you want to return directly to Cambridge or Salisbury from Elliott Island, you could take one very short detour instead. From Vienna continue east on Md. 331 past the entrance ramps to U.S. 50 and take the next right onto Indian Town Road. Where the road bends left keep straight on the unpaved road that runs beside the big pond next to U.S. 50. This pond is too new to have much of a history, but it has habitat attractive to both waterfowl and shorebirds.

A longer detour may be made as you go through Easton. At the 7-Eleven store turn east on Md. 331 from U.S. 50, go 0.5 mile and take the second right on Chilcott Road. In 1.5 miles turn right at the T-junction. In 0.25 mile, the next left, North Dover Road, leads to the **Easton Waste Water Treatment Plant,** one of the best and most accessible shorebird spots in Maryland, and one almost as birder-friendly as its counterpart in Baltimore. If you are punctilious about complying with the ground rules, it promises to remain so.

On Mondays to Fridays (except holidays) you may simply drop in from 7 A.M. to 4 P.M., tell a staff member that you are there to bird, and let them know when you leave. If you want to come on a weekend, call a day or so in advance to let them know. The number is (301) 822-5725, and the key introductory phrase is "I am a D.C. birder." If three or four of you come together, they will leave a gate open for you. If only one or two of you are coming, let them know anyway, and it may be possible to work something out. *Do not trespass under any circumstances.*

Once you are inside, you may go anywhere. From the main entrance gate the second and third roads to the right both lead to the dike between two lagoons, and you can drive around the smaller one, which is managed for shorebirds in summer. The twenty species of shorebirds recorded there one recent summer included all the common species (except Black-bellied Plover, Ruddy Turnstone, and Sanderling), as well as Lesser Golden-Plover, White-rumped and Baird's sandpipers, Long-billed Dowitchers, and Wilson's Phalarope.

In winter and, especially, in migration the waterfowl drop in in ever-changing variety and any species from Tundra Swans to Northern Shovelers, Canvasbacks, and Redheads may be present.

As for songbirds, swallows, particularly Banks, are common in fall. Migrating warblers like the trees along North Dover Road, and Summer Tanagers, Blue Grosbeaks, and Northern and Orchard orioles come to breed. Grasshopper Sparrows nest in the fields nearby.

27 From Salisbury to Broadkill Marsh

Most birders who are not beach-bound themselves stay well clear of the Atlantic resort communities and parks in Maryland and Delaware from Memorial Day through mid-October. The towns are full of people and the parks charge high fees, and the coastal birds are easier to find on Assateague Island or around Bombay Hook and Little Creek.

The most popular time for birders to visit the coast from Ocean City to Broadkill Beach is from November to April, when they have the best hope of finding wintering raptors and the rare northern water birds that sometimes mingle with the more dependable ducks and gulls.

Here is a classic birding run up the Maryland and Delaware coast that can stand on its own or be combined with a pelagic trip out of Ocean City or a day at any of the Delmarva refuges for a satisfying winter weekend.

If you go prepared physically for the coldest winds and psychologically for a total dearth of rarities, you are likely to find much to enjoy. Start early to fit everything in.

The tour begins with a quick visit to the Salisbury dump, variously signposted as the Brick Kiln Road Landfill and the Newland Park Landfill. At the western end of Salisbury, just beyond the 7-Eleven store, turn right at the first traffic light you come to as U.S. 50 curves east, onto Md. 349. Fork right in less than 0.2 mile and right again at 0.6 mile. At 1.6 miles you reach the dump. Follow the trucks through the gate to the right, announcing your purpose to the guard if he is there. The thousands of gulls present will be obvious; the most likely rarities are Lesser Black-backed, Glaucous, and Iceland gulls, among the abundant Ring-billed and Herring gulls. Park well out of the way of the trucks.

Then continue east to the intersection of U.S. 50 and U.S. 113 near the town of Berlin. Take the exit for U.S. 113 south and, exactly 0.5 mile from U.S. 50, park on the shoulder of the ramp. Fifty yards to the

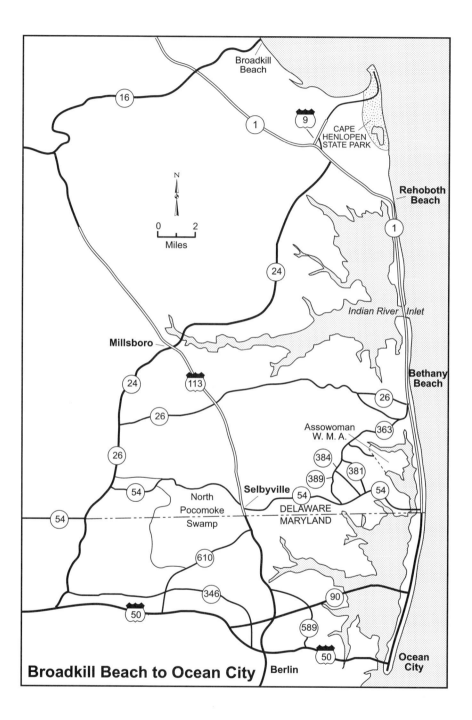

Broadkill
Beach

16

1

9

CAPE
HENLOPEN
STATE PARK

N

0   2
Miles

**Rehoboth
Beach**

1

24

Indian River   Inlet

Millsboro

24

113

26

**Bethany
Beach**

26

363

Assowoman
W. M. A.

26

384

389   381

54

54

Selbyville   54

North
Pocomoke
Swamp

54

DELAWARE
MARYLAND

610

346

90

50

589

50

Ocean
City

Broadkill Beach to Ocean City   Berlin

right are two sewage lagoons separated by a dike. Scope them from the road for ducks and gulls. Then make a U-turn at the next cross-over and return to U.S. 50 (mile 0.0).

At 2.4 miles turn left (north) off U.S. 50 on Md. 589 and, at 2.9 miles, go left again on Griffin Road. Three ponds are on the left before the road bends left at 3.6 miles. If you round the bend, you will be able to study the birds on the far side of the sandbar in the third pond.

If you continue north on Md. 589 0.3 mile beyond Griffin Road, you cross over a creek that sometimes has ducks and gulls right beside the bridge. You can study them safely from Gum Point Road, a right turn just beyond the creek.

The next stop, West Ocean City Pond, is on Golf Course Road, which runs north from the traffic light on U.S. 50 by an Exxon station 3.7 miles east of Md. 589 and just west of the bridge into Ocean City. The pond, 0.4 mile north on the left, is quite large, with inaccessible corners. Do not trespass on adjacent property, no matter what has just swum out of sight.

This pond is noted for its hundreds of Tundra Swans and Canvas-backs, but may easily turn up a dozen other species from herons to gulls, and the other ponds may have quite different populations. It pays to check all of them.

On entering Ocean City, turn right immediately on Philadelphia Avenue, and follow it to the end. Check the water tower for a Pere-grine Falcon—there is almost always one sitting on the lee side in winter, especially in the afternoon. Turn in (only in winter) at the entrance to the Oceanic Motel. The little harbor on the right some-times has loons, grebes, cormorants, and gulls, and at high tide Purple Sandpipers and Ruddy Turnstones may be resting on the rocks and the pier.

Drive on through the motel and straight ahead to the north jetty of Ocean City Inlet. You can scan the inlet, ocean, and south jetty from your car, or park in the lot and walk carefully out along the rocks.

Winter regulars here are Common and Red-throated loons; Horned Grebes; Gannets; Great Cormorants; Brant, Greater Scaup, Oldsquaw, and Red-breasted Mergansers; all three scoters; Purple Sandpipers; Ruddy Turnstones; and Bonaparte's, Herring, Ring-billed, and Great Black-backed gulls. To see and identify them all, a spotting scope is essential. Only a few will be present on any one visit.

The rarities that show up at the inlet most often are Harlequin Ducks, King and Common eiders (usually on the south side of the south jetty), Black-legged Kittiwakes, and Little, Common Black-headed, Glaucous, and Iceland gulls. Expect none of these, but scour the area thoroughly and you may sometimes come up with one or two.

Leaving the inlet, you will be heading north on Baltimore Avenue. At the onion dome of the town hall, turn left on Third Street and follow it to its end at Sinepuxent Bay. You will be at the best vantage point for scanning the "Fourth Street Flats," tidal flats that actually run from Second to Seventh streets.

At low tide hundreds of gulls (including all the inlet species), terns, shorebirds (especially in May), and geese rest here and feed in the adjacent waters. They are easy to study in morning light but become silhouettes on a sunny afternoon. Look for a Gull-billed Tern there in May, Sandwich Terns in late July and August, and a Lesser Black-backed Gull in late September and October.

You can work your way easily enough north along the bay to Ninth Street, at which point condominiums start getting in the way. At that point go back to Philadelphia Avenue, the main road north. If you have time to spare, go out to the ocean beach at every opportunity, especially near the north end of town. The ocean here is often better for loons and grebes than the inlet. Red-necked Grebes; Harlequin Ducks; Lesser Black-backed, Common Black-headed, and Glaucous gulls; and Razorbills and Common Murres have all been reported from this stretch.

At the Delaware state line go left on Del. 54 for 4.0 miles and turn right on Road 381 (mile 0.0). Bear right at 2.1 miles on Road 384 and then follow the signs for Camp Barnes until they are joined by the ones for **Assawoman State Wildlife Area.** At 4.9 miles turn right into the entrance to Assawoman, a beautiful preserve of pine woods, swamps, impoundments, and bay waters laced with firm, sandy roads that should all be investigated. The road to Strawberry Landing, which is closed to vehicles, is particularly recommended for a walk.

Brown-headed Nuthatches and Pine Warblers are resident here, and many ducks and land birds spend the winter. Passerine migration is rich in the spring, and Chuck-will's-widows are abundant then along the roads at dusk. The area is open from 7 A.M. to 8 P.M. Do not go in hunting season except on Sundays.

When you leave Assawoman, turn left, and then keep right at every fork and T-junction (disregarding all side roads). After 6.7 miles you will find yourself at the traffic light on Del. 1, on which you will turn left. (If you do not go to Assawoman, you will reach this same point on the main road north from Ocean City, 6.1 miles above the turnoff onto Del. 54.)

If you are approaching Assawoman from Bethany Beach, go west on Del. 26 for 0.1 mile and go left on Kent Avenue. In 1.4 miles, after crossing a humpbacked bridge, go immediately left at the sign for Camp Barnes on Del. 363. In 3.3 miles go left at the signs for Camp Barnes and Assawoman. (Despite their ambiguous arrows they indicate just one road.) Follow them to the wildlife area.

From the intersection of Del. 26 and Del. 1 go north on Del. 1. In 4.6 miles you will come to the bridge over **Indian River Inlet.** Exit right before the bridge and drive to the far end of the big parking lot. The pumping station offers inlet scanners good protection from winter winds. This inlet is much narrower than Ocean City's, and you can examine both sides of both jetties. It has all the same possibilities for birds, but is seldom as productive. It is worth your while to drive west under the bridge to the trailer park and marina at the end of the road and to check the marshes, sandbars, and open water all around. The best place in Delaware to see oystercatchers and Brown Pelicans, except in the colder months, is the breakwater beyond the little harbor. Red-necked Grebes have been seen here. (At this writing the marina is inaccessible because of construction work, and the replacement of the trailer park with condominiums is rumored. If that happens, access may come to an end.)

If the campground on the way is open but unoccupied, drive along the seawall and survey the inlet and the bay to the west. In summer, terns often hunt in the inlet, especially on an outgoing tide. The marsh just east of the trailer park on the south side of the road is good for Clapper Rails.

To continue north on Del. 1, you must return to the point at which you left it, on the east side of the highway. Cross the bridge and turn immediately left into Indian River Marina. You can pick up information on all the Delaware parks at the state park information office. Drive to the far southeast corner of the parking lot and walk under the bridge out to the beach. There you can check the north side of the jetty and see what is on the beach. This surfers' beach may be closed during nesting season if Piping Plovers set up territories there. Otherwise it is a good place for gulls in winter, terns in summer (Leasts sometimes try to nest here), and seabirds in season.

Then check the inlet, the bay, and the waters between the marina and Burton Island, to the west. A Little Gull may be among the Bonaparte's Gulls. Snow Buntings may be on the sand or around the parking areas. You are welcome to walk out on the docks to look for the Oldsquaws, ruddies, scaup, goldeneyes, and Brant, and to look for an eider among them. If you drive to the northwest corner, where parking for cars and trailers is indicated, you can walk across the causeway to the nature trail on Burton Island.

From the inlet to the town of Dewey Beach is a fast 5.6-mile run, but keep your eyes open. Roadside birds reported along this stretch include an American Bittern, Rough-legged Hawks, a Short-eared Owl on a signpost, a flock of Snow Buntings, and another of Common Redpolls. All the turn-offs to the bay are worth checking. The southernmost, Haven Road, is a loop road at low tide. It is the sort of place that attracts rarities: Wilson's Plover, Sandwich Tern (both in summer), and Common Black-headed Gull have all shown up here, and ducks like it in bad weather. At the second light in Dewey Beach go left into the Ruddertowne shopping center lot for views of Rehoboth Bay. Keep your eyes open for a Peregrine.

When Del. 1 bears left, go straight instead down Bayard Avenue toward Rehoboth Beach. Continue straight to the end of the road, which has become Second Street, where a left turn on Del. 1A will take you back to Del. 1 north of town. A right turn at the lake—which appeals to gulls (including rare ones) as well as to domestic ducks— will lead you in one block to a left turn across the lake. On the other side, go right, and then keep straight and park at the beach beside the end of the boardwalk. In autumn this is a fine place to watch raptors migrating down the coast.

Continue up the beach road and bear left on Surf Avenue (which becomes Zwaanendale Avenue) into Henlopen Acres. Turn left at the T-junction, and stop at the little marina. Brown-headed Nuthatches may still nest in a piling near the steps. Hooded Mergansers like the sheltered water here. On a springtime visit go up Dodds Lane, which becomes Second Street, and turn right at the divided street, Park Avenue. At the end, after the lanes merge and the road curls left, park at the bend and look for migrant warblers. A right turn on Oak Avenue, a left on Third Street, and a right on Lake Avenue will get you back to Del. 1A.

About 3.3 miles north from the junction of Del. 1 and Del. 1A, fork right and follow the signs to the Lewes–Cape May ferry. The open fields in this area are often visited by Brant and Snow Geese. They may also be teeming with blackbird flocks, and with patience and luck you may find Yellow-headed Blackbirds among the Redwings.

Just before a high bridge looms up ahead, turn right at the Boat Factory. Immediately beyond it on the left is a pond that birders regularly check for resident Black-crowned Night Herons and a constantly changing waterfowl population. (If you are here in May, look for Bobolinks in the fields across the road.) Return to the highway, continue to the visitor parking area at the ferry slip, and scan the neighboring waters for loons, grebes, cormorants, and sea ducks. Look for Horned Larks in the grass by the terminal. Then drive on toward Cape Henlopen State Park (free in the winter, but costly for nonresidents from Memorial Day to Labor Day).

Just inside the park, immediately beyond a barracks-like two-story building, a left turn will take you to a parking area beside a long pier. Scan the beach and water on either side, and walk out on the pier as far as you can. The breakwater beyond the pier, especially the tower at the left end, is a fairly reliable resting place in winter for both Great and Double-crested cormorants, and there is a nesting box on the lighthouse for the resident Peregrine Falcon. The pier has been known to shelter eiders and, once, a Western Grebe.

Back on the main park road you will come to the Seaside Nature Center, open all year and well stocked with field guides. If you are lucky, there may be a birder on the staff. Then you pass a huge playing field that is much visited in migration by grass-loving shore-birds and is a nesting ground for Horned Larks. Leave it to your right and take any open road left as far as you can go, beyond the pine-encircled radar tower to a parking area that gives you a view of the whole cape and access to its dunes and beaches. In migration, warblers like the trees and bushes along the road, and from late spring to early fall storm-petrels may be seen offshore from the parking lot.

The dunes may host in winter a Short-eared Owl or two, the Ipswich race of the Savannah Sparrow, Snow Buntings, Lapland Longspurs, and occasionally Common Redpolls and Pine Siskins. The beach, especially at low tide, offers hope of all the birds listed for the Ocean City inlet and flats (except Purple Sandpipers) and the joy of a walk by the sea besides.

The entire cape, with its pines, myrtles, and scrubby vegetation, is worth birding for any wintering species from falcons to shrikes to crossbills. Just before the tollbooth, at a sign directing you to the state park office and information kiosks, a road runs south, past trails into the pine woods and parking lots for observation towers (relics of World War II), to a high overlook next to the sea, a short trail to the beach, and a good exhibit on beach migration.

Retrace your route past the ferry turnoff, over the high bridge, and back to Del. 1, on which you head north for 8.0 miles. At that point the highway crosses Del. 16, the best route back to Washington. Follow the signs for the Bay Bridge.

If you reach Del. 1 before dusk, there are two more stops you can make. About 4.7 miles north of the road to the ferry you will pass little White's Chapel with its ancient graveyard. Just beyond it turn east on unmarked Road 264 and drive 2.5 miles to Oyster Rocks (just a name—no buildings, no rocks) on the banks of Broadkill Creek. The view north to Broadkill Beach and to the great marsh west of it, a part of Prime Hook National Wildlife Refuge, is splendid at any time of year, with herons, shorebirds, and terns present in the warmer months and raptors and waterfowl in winter. You may want to end the day here or go on to Broadkill Road, which winds through the marsh to the beach.

If so, go back to Del. 1, continue north to Del. 16, and turn right. The best place to park and see the entire marsh is by the bridge at 2.8 miles from Del. 1. (A better stop for birding in full daylight is two-thirds of a mile farther on, where ponds and flats spread out on both sides of the road.)

Noted in summer for its Black Rails (to be sought only from the road), in winter Broadkill Marsh is quartered at sunset by Northern

Harriers, Rough-legged Hawks, and Short-eared Owls, while skeins of Canada Geese snake in thin, black lines against the reddening western sky, their calls mixing with the hoots of the Great Horned Owl in the nearby woods. It is a perfect place to end the day.

North Pocomoke Swamp, Deal Island, and a Winter Spectacle **28**

Cypress Swamp Conservation Area, a sanctuary on the Delaware-Maryland line owned by Delaware Wildlands, Inc., is widely known as **North Pocomoke Swamp** to East Coast birders, who converge on the most accessible stretch of this beautiful swamp during the first three weekends of May.

The drawing power of these dates is the excellent chance of seeing a variety of migrant songbirds in addition to the nesting species that are the main attraction. If you do not mind missing the migrants, which can be found as readily elsewhere, come in late April or from late May to early July, when you can study in peace the nesters and perhaps their endearingly curious fledgling young.

If you arrive before sunrise, Barred Owls may still be calling. Nonpasserines are likely to include Green-backed Heron, Red-shouldered Hawk, Yellow-billed Cuckoo, and Pileated Woodpecker. Great Crested and Acadian flycatchers, Eastern Wood-Pewee, Wood Thrush, Blue-gray Gnatcatcher, and White-eyed, Yellow-throated, and Red-eyed vireos are near-certainties.

It is the warblers, though, that are the magnet: Black-and-white, Prothonotary, Worm-eating, Parula, Yellow-throated, Pine, Ovenbird, Louisiana Waterthrush, Kentucky, Hooded, and Redstart.

The easiest route to the swamp from the west is from U.S. 13, where it crosses the Maryland-Delaware state line. Go east at that point (mile 0.0) on Del. 54 for 9.1 miles. When Del. 54 turns north, continue east on Bethel Road, bearing right at the fork at 11.2 miles. At 11.9 miles turn left on Sheppard's Crossing Road.

Shortly after entering the swamp you will cross the Pocomoke River at 12.6 miles. Park near it and do a little birding on either side of the road. Then continue to 13.1 miles and turn left on an unpaved road. When you cross the next bridge, at 13.4 miles, park somewhere safe, either on the road ahead (north) or the one to the right (east). Bird along both roads and explore any little trails you see.

To reach the same entrance from U.S. 50 turn north on County Road 610 about 17 miles east of Salisbury (mile 0.0). After crossing Md. 346 in 0.5 mile continue to the north end of Dale Cemetery at 1.2 miles and fork left on Sheppard's Crossing Road. Then keep right at all forks until you reach a T-junction at 2.8 miles. Turn left. At 4.6 miles turn right on the unpaved road into the swamp.

Most birders spend their time in this part of the swamp. If you continue another 0.7 mile beyond the bridge you will be out of the woods and out of Maryland. When you enter the next woods at 1.5 miles you will have a long stretch of swamp, (probably all to yourself), a currently neglected trail by the old Delaware Wildlands sign, and all the same birds.

Emerging from these woods at 2.3 miles, you meet Del. 54 at a T-junction at 2.8 miles. Turn right if you are going to the coast and drive once more through the swamp, birding at will along the highway. (A left turn, of course, will return you eventually to U.S. 13.)

At 8.5 miles fork left on Del. 54 if you are heading east or north; otherwise continue straight until you meet U.S. 113 at 9.1 miles.

The reverse route from the junction of U.S. 113 and Del. 54 (mile 0.0) in Selbyville, Delaware, is even easier. Drive west on Del. 54 through the swamp and go left at 6.2 miles just after the curve sign. This is the unpaved road that first winds through the Delaware swamp and then runs through the Maryland swamp.

You will reach U.S. 50 by leaving the swamp at the south end. Turn left on Sheppard's Crossing Road (mile 0.0) and turn right at 1.8 miles. Continue south to Md. 610 at 3.3 miles and bear right; you will run into U.S. 50 at 4.5 miles.

Along all the back roads in this area be on the alert for Blue Grosbeaks and Orchard Orioles, which are common to abundant along the edges of the farm fields, and for the Horned Larks that breed in the fields.

All the same species can be found in Pocomoke State Park, 3.5 miles southwest of Snow Hill, Maryland, on U.S. 113, and in the adjacent Pocomoke State Forest to the west. Investigate the trails that run back into the woods on the north side of the highway in the first two or three miles west of the park entrance.

While many birders combine a morning at North Pocomoke Swamp with an afternoon on Assateague or around Ocean City, the rest of the day can also be spent profitably on an excursion to **Deal Island Wildlife Management Area,** home to nesting Least Bitterns, Pied-billed Grebes, Gadwall, Blue-winged Teal, Bald Eagles, Northern Harriers, Ospreys, Common Moorhens, American Coots, Black-necked Stilts, Marsh and Sedge wrens, and Sharp-tailed and Seaside sparrows.

It is intensively used as a feeding area by all the herons and ibis nesting in nearby colonies and by the Forster's Terns that breed on the little marsh islands offshore in Chesapeake Bay. Normally, water levels in spring are too high to attract any shorebirds but yellowlegs; both species of dowitchers and White-rumped Sandpipers may turn up in the fall, however, and Common Snipe winter in the marsh. In the rare years when the impoundment is drained, thousands of shorebirds pour in.

In winter the area shelters large numbers of dabbling ducks and is a popular spot to look for Eurasian Wigeon among the American Wigeon. In addition to the Harriers and Bald Eagles, likely raptors include Red-tailed Hawks, Peregrines, Merlins, and Short-eared Owls. Rough-legged Hawks are often astonishingly abundant, with a dozen or more in view at once.

Deal Island is at its best at the end of the day, when it is alive with marsh birds. To reach the area, take U.S. 13 south from Salisbury to Princess Anne and turn right on Md. 363. Drive 9.5 miles and fork left at the Deal Island WMA sign. Keep right where a road branches left after 1.2 miles and continue to the chain across your path and a big green dumpster at 1.4 miles. This is the best spot to see Black-necked Stilts, Soras, and Least Bitterns in summer and Sedge Wrens in late fall.

Go back to Md. 363 and continue west 2.2 miles to Dame's Quarter, where you take the first possible left (mile 0.0), on Riley Roberts Road, which goes through a small settlement then gradually deteriorates, but is always navigable. You might park by the chained-off road on the left, at 1.7 miles, and walk as far as you like along the north side of the impoundment. Then continue to the end of the road by the Manokin River at 3.3 miles; you will have driven along the west side of the impoundment. A short walk out the dike from here brings you to the best spot for Short-eared Owls and American Bitterns. In fact, it is a good area to see all the herons, including a large group of Black-crowned Night-Herons that usually takes off at dusk. If you are in the mood to hike eight or nine miles you can circumnavigate the impoundment on foot.

When you leave the wildlife management area, do not miss the opportunity to see the quintessential watermen's communities along the highway to the west. After passing through Chance and Deal Island, the road ends in 6.7 miles at Wenona, where you can look out to Little Deal Island and over Tangier Sound. In November dozens of loons can be seen from here (hundreds of them through a scope on a clear day), and perhaps gannets plunging into the water far offshore.

If you while away the hours until nightfall, the marshes can be full of the sounds of rails. Virginias and Soras are the most common, though there are pockets with many Clappers, especially near the sound. Blacks are widely scattered and less numerous. At least a few

King Rails are probably here, as well as possible King × Clapper hybrids. You are unlikely to see any rails without the aid of a tape recorder. You will have just as much chance to bring them to you if you stay on the road as you would if you went into their fragile habitat.

Before you set out, be aware that the marshes of Somerset County can be infested, day or night, with unbearable numbers of mosquitoes. An ample supply of concentrated insect repellent may be as crucial to your pleasure there as binoculars.

One of the highlights of an autumn or winter visit to this corner of the Maryland Eastern Shore is the sight of a remarkable concentration of Bald Eagles (often forty or more) that roost along the Pocomoke River just west of Pokomoke City. If you are in the area at dawn you can see them leaving the roost, but later in the day they feed in nearby farm fields or sit in adjacent trees, all in a compact area that needs no distance measurements. From U.S. 13, 11 miles south of Princess Anne, turn west on Road 667, Rehobeth Road. If you come at dawn, go about 2 miles and turn left down unpaved Powell Wharf Road until you have a clear horizon to the left. The eagles leaving the roost often fly low over the road.

If you come later in the day, pass Powell Wharf Road and turn right in a half-mile on Gordy Road, left on Mennonite Church Road, and right on Elmo Dryden Road. In a half-mile you will pass the several buildings of the Dryden Farm (which includes a hairdresser's shop). This is likely to be the best place to see many eagles at once, in the trees or on the ground. If they are not there, explore the neighborhood, or ask questions. At the end of Elmo Dryden Road, a left turn on Barnes Road and another on Road 667 will take you right back to U.S. 13.

29 The Northern Half of Assateague

As this book was in press, a winter storm caused many changes to the north end of Assateague Island and made the original text for this chapter obsolete. In response to an urgent plea for help, Mark Hoffman, whose experience in birding Assateague is unsurpassed, generously drafted the replacement that follows.

Each fall the Maryland portion of Assateague Island serves as a mecca for birders. The attractions include waves of migrant land birds, western vagrants, and an impressive falcon migration. The cumulative list—including such species as Sage Thrasher, Vermilion Flycatcher, and Smith's Longspur—is enough to tempt even the most lethargic seeker of rarities. Fall is not the only time to visit the island, however, and trips here can be fruitful throughout the year.

The island is reached by taking Md. 611 south from U.S. 60, 0.8 mile west of the west end of the bridge into Ocean City. You may wish to stop at the Visitor Center for Assateague Island National Seashore to obtain an area map or view the displays.

Just before crossing the high bridge, pull into the boat launching area on the north side of the road. From here it is possible to scan Sinepuxent Bay for loons, grebes, Brant, and diving ducks during the winter months. Walk out onto the fishing pier to observe a small sand spit just south of the bridge near the mainland.

Then cross the bridge onto the island. At the fork, go straight ahead to the day-use parking lot of Assateague Island State Park. During the summer a $2-per-person entrance fee may be charged. Park in the northwest corner of the large lot.

In fall the thickets and pine woods around the lot can be excellent for migrating land birds. The best fallouts of autumn migrants occur immediately after the passage of a cold front. During the first few hours of the day, migrating warblers and other land birds can be seen flying in from the ocean and diving into the nearest thick vegetation. Although many birds will have to be left unidentified, this activity can be exciting to experience. Moreover, the possibility of western vagrants keeps birding interesting, with Western Kingbird, Dick-cissel, Lark Sparrow, and Clay-colored Sparrow of almost annual occurrence.

Try the dirt road leading north from the entrance road just west of the tollbooth (marked by a gate and an "Authorized Vehicles Only" sign). Then return to the entrance road and walk west along it to explore the bayberry and wax myrtle thickets to the north and south of the causeway. In fall these thickets are among the best areas for lingering migrant warblers. Explore the marshes, particularly to the north of the road, for Sharp-tailed Swallows (all year) and Seaside Sparrows (in summer). You may find it worthwhile to walk south from the day-use parking lot and explore the areas around the State Park bathhouses; the various camping loops provide a convenient footpath.

For the adventurous and energetic birder, the north end of Assateague provides some of the best shorebird habitat in the state. Although it requires a 3-mile walk to the first flats and a strenuous 6.5-mile hike to the south side of Ocean City inlet, this trip can prove well worth the effort. It is best planned for when low tide is scheduled for 10 to 11 A.M. The tidal flats are on the west, and at any time of year it is best to be looking toward the bay in the morning and toward the ocean in the afternoon. During recent summers, access to the north end has been restricted to the ocean beach from April to August to protect the endangered Piping Plovers, which nest on the shell-strewn sand flats. Check at the National Seashore Visitor Center regarding limitations on access to this area at the time you want to visit. If you go in summer, wear a hat and take plenty to drink, insect repellant, and sunscreen. Be prepared for wet feet in any season.

Recent storms have played havoc with the topography of Assateague, and specific landmarks are virtually nonexistent. Walk north from the day-use parking lot of the State Park for about two miles. Assuming unrestricted access, cross west to the bay side once the pine woods end on the far shore and continue north along the edge of the bay. You will soon come to a stretch of tidal flats, sand beach, and marsh. Among the first flats is a very large one with marshy islands bordering its west side, just south of the first large housing development on the opposite shore. During southbound migration, Marbled and Hudsonian godwits occur here, along with more common species. Continue north, checking the flats for shorebirds and the shallow bays for waterfowl, as far as the inlet, or turn back whenever your endurance starts to run out. The return is best made along the ocean beach.

If you skip exploration of the north end, leave the State Park day-use area and turn south along Bayberry Drive to the public-use facilities of Assateague Island National Seashore. At 2.4 miles you will reach an entrance station, where you are charged a $3 entrance fee (or $10 for an annual pass), although in the off season the station is not staffed. A large parking lot adjacent to the ranger station and campground office is immediately on your left. From the parking lot a wooden walkway provides access to the ocean beach and to a convenient spot to scan for Northern Gannets, Brown Pelicans, cormorants, scoters, and seabirds like jaegers and shearwaters.

As you leave this area, Bayside Drive, leading to and beyond Bayside Campground, is just down the road on the right. Turn in here and proceed to the parking lot for the Life of the Marsh Trail on your left. The entire Bayside Campground area is excellent for migrant land birds in fall. The brush and scrub areas frequently host Orange-crowned Warblers and Clay-colored Sparrows, and the large willow trees on the north and south sides of the campground are good for Philadelphia Vireos. The area is best explored from the Life of the Marsh Trail and the three camping loops.

At the end of Bayside Drive a picnic area provides a good place to scan the bay for water birds. This point is also excellent for fall migrants. After the passage of a cold front, birds will be attempting to work their way back to the mainland, and they become concentrated here. Loose flocks of warblers rise from the scrub and start to take off across Sinepuxent Bay, only to return and dive back into the brush.

Another productive area for migrants is the Life of the Forest Trail, on the right of Bayberry Drive, 0.5 mile south of Bayside Drive.

In another 0.4 mile, the road for two-wheel-drive vehicles ends at the parking lots for the Life of the Dunes Trail and the South Ocean Beach. The southwest corner of the trail borders a large pine woods that can be wonderful in migration. During spring and fall be sure to check the deciduous trees among the pines for warblers and other migrants.

Another trip for the hiker, and a less exhausting one than the trip to the north tip of the island, is a visit to the Hungerford flats. From the South Ocean Beach parking lot, walk over to the entrance to the Off-Road Zone. Follow the ORV (off-road vehicle) road south. You will pass a road leading to the Tingles Island hike-in camping area, which may be worth a short side trip. Continue south to the first set of power lines that head west from the main road. Follow the sand road paralleling the power lines toward the High Winds Gun Club. After you go through an area of thick bayberry scrub, the vast Hungerford flats will appear to the south. This is the prime shorebird area in the southern part of the National Seashore. It is possible to walk along the road and dike to the west end of the flats and head south to explore the area. A better strategy, however, is to return to the main road and continue south to Dune Crossing Number Five. From here follow the road west, first to a small parking lot and then to a channel that borders the southeast corner of the flats. First go south at the channel to check the north end of the Little Levels, often a good shorebird area. Then head back north and cross the channel where possible—at high tide hip boots (or wet feet!) will be required. The Hungerford flats will now be to the northwest and in good light at any time of day. It is normally most productive to walk north along the west side of the flats until you eventually reach the High Winds Road. Follow this road east back to the main road and then go north to return to your car. Walking time from your car to the north end of the flats is approximately one hour.

Note: From late May through September, mosquitoes can be horrendous on Assateague. Be prepared to coat yourself entirely with insect repellant unless you plan to visit only the ocean beach.

Chincoteague 30

Chincoteague National Wildlife Refuge may be the best place on the Atlantic coast to see a succession and combination of virtually all the

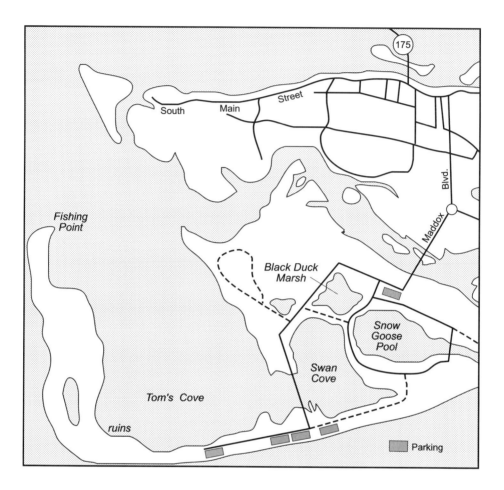

eastern loons, grebes, cormorants, herons, ibis, swans, geese, ducks, falcons, shorebirds, gulls, terns, and skimmers, along with land birds ranging from resident Fish Crows, Brown-headed Nuthatches, Pine Warblers, and Boat-tailed Grackles to breeding Seaside Sparrows, and wintering Ipswich (Savannah) Sparrows and, sometimes, Snow Buntings.

If you enjoy a world of marsh, pine woods, wild beach, and open ocean, it is a beautiful and fascinating place to visit and bird all year-round, with abundant and convenient tourist accommodations. It is an easy area to bird by car, but gives its greatest rewards to the birder with the time and energy to explore it on foot.

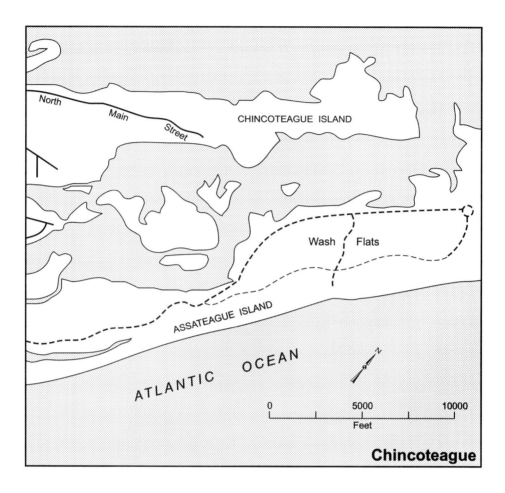

North

Main
Street

CHINCOTEAGUE ISLAND

Wash Flats

ASSATEAGUE ISLAND

ATLANTIC OCEAN

0   5000   10000
Feet

Chincoteague

Allow a solid three hours to drive from the Capital Beltway, I-95/495, via U.S. 50 east to U.S. 13 in Salisbury, U.S. 13 south to Va. 175, and Va. 175 east to Chincoteague. If you like back roads, an alternate route from the U.S. 13 bypass around Salisbury is Md. 12 to the Virginia line, where it becomes Va. 679. This road intersects Va. 175 east of U.S. 13. Turn left.

If you arrive when the tidal flats are exposed, begin your birding along the causeway from the mainland, just before you cross the second bridge. A short spur road on the right, 1.4 miles from the west end of the causeway, provides close views of an oysterbed, where you should find at least two—perhaps a hundred—American

Oystercatchers, as well as other shorebirds, egrets, Laughing Gulls, and Forster's Terns in summer, and Brant, gulls, and perhaps Horned Grebes, Double-crested Cormorants, and diving ducks in winter.

Immediately east of the bridge, pull carefully over to the shoulder on the north side of the road and park. The extensive mud flat attracts not only Brant, gulls, terns, and oystercatchers, but also Black-bellied Plover, Willets, Short-billed Dowitchers, Dunlin, and "peeps." The same shorebirds may be found at low tide on the flats beside the north-side pull-off 0.3 mile farther east.

Drive on through the village of Chincoteague, which is on Chincoteague Island, to Chincoteague Refuge, the Virginia end of Assateague Island. (Part of the refuge is managed by the National Park Service under the name Assateague National Seashore.) Follow the signs to Assateague and you will be on the refuge. You will have to have a National Park Pass or a "duck stamp," or pay a fee. Stop first at the Refuge Information Center, and pick up the available handouts, including a bird list. Check the wildlife sighting list, usually hung on a nail outside the door. Look and listen for Brown-headed Nuthatches and Pine Warblers around the parking lot. (Brown-headed Nuthatches are always somewhere in the pines, but they wander widely.)

If you arrive after 3 P.M., you can take your car around the 3-mile Wildlife Drive (open only to pedestrians and bicyclists earlier in the day), which loops around Snow Goose Pool and through pine woods. When water levels are high, usually November to April, it is full of waterfowl; as it dries up from May or June to July or August, it is a mecca for feeding egrets and ibis and for resting gulls, terns, and skimmers. When mud flats appear, it becomes prime shorebird habitat, first for yellowlegs, Short-billed Dowitchers, Stilt Sandpipers, and an occasional Ruff, then for Pectoral Sandpipers and peeps. In dry years (at least three out of five in recent years), Snow Goose Pool dries up entirely and attracts "grasspipers" like Lesser Golden-Plover and Buff-breasted Sandpipers. Vegetation in this impoundment grew rapidly in the 1980s, and its use by shorebirds in summer and fall plummeted. Peeps and yellowlegs still use the ditches around the edges as long as there is mud and a little water, and Pectoral Sandpipers and other grasspipers can be found from August to October. (You may walk out into the impoundments if they are dry and if you find a way to get beyond the ditches.) The shorebirds are magnets for migrating Peregrines and Merlins in September and October. The pine woods halfway around may shelter a flock of Brown-headed Nuthatches and are sometimes swarming with warblers in migration. A morning walk around the Wildlife Drive allows you to see in good light areas that are backlit in the afternoon.

Back on the main road, keep your eyes open for White-tailed and Sika deer, feral ponies, and cars stopping suddenly to look at them.

Black Duck Marsh, the first impoundment on the left, is especially attractive to Snow Geese, Green-winged and Blue-winged teal, and other dabbling ducks, and is a good spot to look for Eurasian Wigeon in October and November. In summer it dries up earlier than the other impoundments and is often the best area for shorebirds in May and egrets and ibis in June and July. In these months, particularly at dawn, you may see Gull-billed and Forster's terns hawking for insects over the shallow water.

A little before you reach the marked entrance to the Woodland Trail on the right there is an unmarked pull-off on the left from which you have the best views of the southwest part of Swan Cove. You will need a good telescope and afternoon light. Waterfowl and herons often shelter from northerly winds behind the little islands, and terns, including Black Terns, and shorebirds rest on the mud flats. In late summer and early fall, Avocets and Hudsonian and Marbled godwits feed in the shallows.

The Woodland Trail is so lovely that it is a pity that insects make it unbearable for most of the summer. If you get up early enough to be the first to take this 1.6-mile loop you are quite likely to meet a Sika Deer face to face, get a close view of a Delmarva Fox Squirrel, and stare down a herd of ponies blocking your path. It is the best area on the refuge for migrant and wintering land birds.

Along the trail are two areas once prepared as parking lots. If you go around the loop clockwise, you will find a path running off to the left just beyond the first of these lots (nearly halfway around). Follow it almost no distance to the beach on the east side of Tom's Cove, which you will probably have to yourself. A walk toward the point on your right can be a delight, particularly in spring migration, for both water birds and land birds.

The causeway between the east half of Swan Cove and the north end of Tom's Cove, a tidal bay, is usually worth a number of stops. Look for Mute and Tundra swans, Canada Geese, Ruddy Ducks, and Pied-billed Grebes to the left, and Common Loons, Horned Grebes, Brant, and Red-breasted Mergansers to the right. You may see Double-crested Cormorants and Ospreys on the pilings and hunter's blinds; American Oystercatchers, Ruddy Turnstones, and Black-bellied Plover on the oyster shells and tidal flats; Whimbrel, Clapper Rails, and Seaside Sparrows in the salt marsh; and yellowlegs, dowitchers, peeps, gulls, and terns anywhere. Especially near dusk in summer look for Black- and Yellow-crowned Night-Herons and Black Skimmers along the ditches.

The road to the beach ends at a small circle. Go left to the Tom's Cove Visitor Center and to parking lot #1, which is an excellent spot to study and photograph gulls, mainly Laughing and Ring-billed, in all their plumages, at close range. The bicycle trail that leaves the parking lot from the northwest corner gives good views, especially in morning light, of the birds in the northeast corner of Swan Cove. It then leads past ponds and through woods to the Wildlife Drive. You may walk out into the impoundment if it is dry enough.

Three other parking lots lie south of the circle, and the northern-most of them also attracts gulls. The paved road that used to extend 2 miles down the beach behind the dunes was buried in sand in the winter of 1981–82 and has never been replaced. Forecasts based on studies of beach dynamics predict continuing instability and erosion along the thin neck of land leading to the south end of Asssateague.

Most of the south end of the island is closed from March to August to protect the nesting Piping Plovers, and the long walk to Fishing Point at the tip is not very rewarding for birds in autumn and winter. The area of greatest interest in spring lies between parking lot #4 and the south end of the washover area, on the Tom's Cove side. From mid-May to early June, horseshoe crabs lay their eggs on the beach, and feeding Sanderlings, Red Knots, Ruddy Turnstones, Semipal-mated Sandpipers, Dunlins, Short-billed Dowitchers, and Laughing Gulls gorge on them, especially as the tide begins to fall. You may see American Oystercatchers and Piping Plovers in the area. The cove beach is much less interesting the rest of the year, though terns and skimmers often loaf at the south end of the overwash in late summer and fall, and the loons, grebes, and sea ducks in the cove itself will keep you entertained from fall to spring. The ocean beach is always worth checking; in tourist season do it early in the morning before the sunbathers and many of the fishermen are up. Whimbrels, Sanderlings, and Willets may be out in the surf or up on the beach, gulls may be loafing, and terns may be plunge-diving into the sea.

In September and October, the wires, bushes, and sandy patches along the remains of the paved road south of the overwash are heavily used by migrant passerines, especially swallows, flycatchers, orioles, and sparrows. Check them carefully for Western Kingbirds, Clay-colored Sparrows, and Lark Sparrows. Continue beyond the end of the paving as far as the old Coast Guard dock (though you may not walk on the dock itself), and come back along the cove past the ruins of the old fish cannery.

A sea watch is best mounted up the beach, near the parking lots. Birds are most likely to be seen from October to May, or after major storms. Look for loons, gannets, scoters, Red-breasted Mergansers, Oldsquaw, and, under ideal circumstances (stormy weather and strong easterly winds), shearwaters and jaegers. In September and October try a hawk watch, especially for accipiters, Ospreys, harriers, and falcons.

The ultimate Chincoteague bird walk is the 14-mile round trip to the top of the Wash Flats, one that can be shortened by 3.5, 4.5, or 7 miles if you take the Safari Wagon (reservations at the refuge information center) to the south end of the flats, the west end of the crossdike, or all the way to the end of the service road. In winter the Safari Wagon does not run and you have to walk the whole distance to see the thousands of waterfowl that fill the 3-mile-long divided impoundment.

In any case, it is a very long trip and it is worth your while to acquire current information about the abundance and variety of birds present before you set out. The Wash Flats can be the most exciting place in the refuge, depending on water levels. If the flats are partially flooded, a July or August day can produce most of the long-legged waders, shorebirds, and gull and tern species on the refuge list. (In recent years, water levels have been lowered in spring to provide nesting habitat for Piping Plovers, and summer rains have not replaced the water lost to evaporation. The impoundments are now usually at their best in May and again in fall.)

The flats on a day in September or October are likely to provide you with more Peregrine Falcons than any other site on the East Coast. It is a reliable spot for Buff-breasted Sandpipers in late August and September, and it is known for such one-shot rarities as a Mountain Plover, a Long-billed Curlew, a Pomarine Jaeger, and an Elegant Tern.

If you make a day of it, starting at dawn and taking a picnic lunch and as much to drink as you can bear to carry, you should find the trip enormously rewarding—but skip it if the forecast is for sunny skies and temperatures over 90° or if mosquitoes and green flies are rampant. If it is good hiking weather, the most interesting route is to go one way along the service road and the other along the beach and the east side of the Wash Flats, choosing the beach route to coincide with the hours nearest low tide. This is possible only from September to February because of Piping Plover protection, and worthwhile only in September and October.

Assuming low tide is in the morning, park at the Refuge Information Center, walk left up the Wildlife Drive, and go through the stile next to the gate at the north corner of the drive. You are now on the service road. After about 2.3 miles, take the second road right, which goes across "D" Dike to the beach. Walk left up the beach about 2.5 miles until just after you pass kilometer post 7 and then cross the dunes and climb over the fence at the east end of the crossdike. Not far ahead you can cross the ditch on the right side of the road. Zigzag north, surveying both the water's edge to the west and the grassy flats to the east, the latter being where Buff-breasted and Baird's sandpipers and Lesser Golden-Plover are most likely to occur.

Continue past the north end of the flats to the road that winds west through the woods to the turnaround at the end of the service road.

The latter runs south along the west side of the Wash Flats, giving excellent views of bayside tidal flats en route. After you pass the gate at the south boundary of the Wash Flats, you can either follow the fence east over to the beach again or continue south along the road back to the Wildlife Drive. The route is easily reversed; you need only to know that "D" Dike is the first vehicle crossover south of the Wash Flats. If you don't have the energy to go to the top of the flats, walk partway up the north impoundment, or make the crossdike your northern limit; you will save 5 miles but miss some good birds.

When you leave the refuge, several other stops are worth making. The easiest place to see Clapper Rails is along the road just west of the bridge between Chincoteague Island and Assateague Island, at the edge of the marsh on either side. They are most likely to be seen, spring to fall, when low tide coincides with dawn and dusk. Lie in wait for them at the end of the tidal gut on the north side of the road.

Main Street, which runs along the west side of Chincoteague Island, is worth cruising in the morning or on an overcast afternoon if the traffic is not too heavy. At the south end is a marshy pond or two where Yellow-crowned Night-Herons sometimes fish from spring to fall. The porch of the Crab House Restaurant provides good views of Chincoteague Channel and the tidal flats beyond, where Brant, gulls, and shorebirds can be studied as they feed or rest.

31 The Virginia Eastern Shore

The Eastern Shore of Virginia south of Chincoteague offers minor and major birding pleasures that rival those of the refuge itself. Two of them require investments of funds and advance planning, both of which are fully justified.

Mileages at points on U.S. 13 cited below are the distances, without the recommended detours, from the Maryland-Virginia line (mile 0.0).

The **Saxis marshes** lie west of Temperanceville (7.8 miles). Turn right off U.S. 13 on Road 695. After 8.6 miles, turn left on Road 788, drive south 0.5 mile, and park on the shoulder. The broomsedge in the marsh on both sides of the road has been a nesting area for Sedge Wrens in summer; try the hours close to dawn. Beyond Road 788, Road 695 runs out into open marsh, where you can expect Clapper and Virginia rails and Seaside Sparrows. Rough-legged Hawks are commonly present in winter. Black Rails have been found here, but not regularly. If you are here by day, go on to the end of the road in Saxis and observe the ways of this single-minded crabbing village. (Well, almost single-minded—the owner of the fourth house on the right before the end of the road has twelve Purple Martin houses on the property!)

The **Tyson lawns.** At 8.7 miles, U.S. 13 passes the Tyson chicken processing plant. Its extensive lawns can be scanned from the highway or from a road along the south side of the property. Gulls are always present, and Lesser Golden- and Black-bellied plovers, and Upland, Baird's, and Buff-breasted sandpipers have been seen there in fall. When there has been so much rain that all the impoundments in Chincoteague are full, migrating yellowlegs, peeps, and Pectoral Sandpipers also find their way here to the rainpools.

The farm fields between Accomac and Quinby warrant a detour to the east to look for shorebirds, but only in a wet spring. Leave U.S. 13 (at 20.7 miles) on Business 13 into Accomac, and turn left in 2.0 miles on Road 605. After 3.8 miles, turn left on Road 647, a loop road that returns to Road 605 in 3.4 miles in Locustville. If rain pools are standing in the bare fields, many hundreds of shorebirds—mostly Black-bellied Plover, Semipalmated Plover, Ruddy Turnstone, Short-billed Dowitcher, Dunlin, and Semipalmated Sandpiper—will all be feeding in the soft earth. (A sometimes worthwhile detour off Road 647 1.4 miles after you turn onto it is a left turn on Road 787, which takes you by way of a gravel road to Burton's Shore, a public beach and boat ramp on Flood's Bay.)

The same conditions may exist here and there along the road as far south as Quinby, 8.4 miles from Locustville. At that point take Va. 182 west 3.8 miles back to Painter on U.S. 13 (34.4 miles). En route you will cross the beautiful marshes of the Machipongo River, worth a detour at any time of year.

The **Virginia Coast Reserve.** The high point of a trip to this part of Delmarva should be, at least once, an expedition to one or more of the barrier islands, most of which are owned and managed as the Virginia Coast Reserve by The Nature Conservancy. If you do not have your own boat, you may be able to rent one in Wachapreague or negotiate for transportation with a waterman at one of the harbors.

The easiest and most rewarding solution is to take one of the boat trips run by The Nature Conservancy, scheduled to coincide with peak migration periods. The boat leaves from one of the waterfront villages and goes out through the vast marshes to one of the islands. The boat ride is followed by a hike along the barrier island. You will be informed, entertained, and relieved of all logistical problems. You will need to bring lunch, and you should take a scope if you have one, as well as a hat, sunscreen, and insect repellant. Be prepared for wet feet. Trips typically depart about 8:30 A.M. and return in mid- to late afternoon. At this writing most of them cost $30 or less. An excellent fact sheet on the tours and a natural history guide for the islands, including a bird list, are available. To acquire more details or to make reservations, write or call: The Virginia Coast Reserve, Brownsville, Nassawadox, VA 23413; telephone (804) 442-3049. Book early.

If you have never had the opportunity to visit a barrier island inhabited only by thousands of nesting or migrating birds, virtually undisturbed by human beings, do not miss this one. On the way you may see, on a spring tour, Brown Pelicans; Great Blue, Green-backed, Tricolored, and Little Blue herons; Great and Snowy egrets; perhaps both night-herons; a variety of migrating shorebirds as well as the resident American Oystercatchers and breeding Willets; at least three species of gulls; Royal, Common, Forster's, and perhaps Sandwich terns; and Black Skimmers. From fall to spring, loons, Horned Grebes, Brant, Snow Geese, Black Ducks, bay ducks of several kinds, scoters, mergansers, Peregrine Falcons and Merlins, and many wintering shorebirds are possible. Depending on when you go, you may see huge flocks of Red Knots massed along a peaty shore, herons on their nests in the marsh elder bushes, courting oystercatchers, or gannets plunging offshore. Rarities on the 260-species checklist include White Pelican, Magnificent Frigatebird, Anhinga, White Ibis, Long-billed Curlew, and Roseate Tern; but you need no rarities to make this an exhilarating experience.

Plantation Creek is a detour toward Chesapeake Bay a short distance south of the Cape Charles intersection. Turn west on Road 644 at the sign for "Custis Tomb." The tomb is that of relatives of Nellie Custis, and it is 2.2 miles down the road, within a housing development called Virginia's Chesapeake Shores. If you continue 0.2 mile farther, you come to an unpaved overgrown parking lot where you may leave your car. Walk toward the water and take the elevated boardwalk that leads in 0.2 mile to a gazebo with a fine view where the creek meets the bay. There are good mud flats at low tide, and birders are welcome on this private property.

Eastern Shore National Wildlife Refuge (ESNWR). At the south tip of Delmarva, this new refuge lies on the east side of U.S. 13. At this writing, facilities for birders are minimal. Part of the refuge used to be a military reservation, and there is still much open space that is not very productive ornithologically. The short nature trail is not a compelling reason to stop for much of the year, and most of the prime birding habitat is closed to casual visitors. Rarities in recent autumns here, however, have included Say's Phoebe, Western Kingbird, and Le Conte's Sparrow, and the old fortification along the nature trail is a high vantage point for a hawk watch. The jewel of the area is Fisherman Island, a refuge in its own right, under ESNWR's jurisdiction, but it is open only for group tours. (The Chesapeake Bay

Bridge-Tunnel crosses the island, but stopping there is absolutely forbidden.) The refuge staff welcomes ad hoc groups, provided arrangements are made far enough in advance, and a visit to Fisherman Island is recommended. It is an outstanding nesting area for herons and egrets, ibis, and gulls and terns, but among the breeders are two threatened species, Brown Pelican and Piping Plover, and the only White Ibis in Virginia. To protect the nesters, no visits at all to the island are allowed between April 1 and September 1. There is excellent habitat for migrating and wintering shorebirds, however, and most of the raptors moving down the East Coast pass over this point. A visit in late September and early October could be very rewarding.

As the facilities at ESNWR are developed, it is likely to become a site well worth visiting. The area is included in the count circle of the Cape Charles Christmas Bird Count, which consistently tops the Virginia Christmas Bird Count totals, and a recently defunct banding station nearby proved the mass movement of songbirds through the neighborhood in autumn. Drop in and find out what is going on.

The **Chesapeake Bay Bridge-Tunnel.** To reach Norfolk and points south from Delmarva, the only direct route is via the 17-mile highway over and under Chesapeake Bay. Whether you want to end up on the south side of the bay or not, a birding expedition on the bridge-tunnel can be worth every penny of the costly toll, except, perhaps, in calm midsummer weather. At least ten days in advance (if you can manage it), write a polite letter to: Executive Director, Chesapeake Bay Bridge and Tunnel District, Post Office Box 111, Cape Charles, VA 23310. If you have not left time for an exchange of letters, call the north office, if you are heading south, at (804) 331-2960, or the south office, if you are heading north, at (804) 624-3511, as far in advance as possible.

Request permission to stop on the bridge-tunnel islands "to observe sea birds and waterfowl." Permission for the entire calendar year should come promptly in the mail, along with certain restrictions (no children in the car, for example), enabling you to stop, park, and circumnavigate on foot the three northern islands at the ends of the two tunnels. If there is not enough time for permission to reach you by mail, request that your letter be left at the office at the appropriate end of the bridge-tunnel. Show your letter to the collector at the tollbooth (69.3 miles), and keep it handy for the patrol officers, who will leave you alone or treat you kindly if you follow the rules.

If you do not obtain this written permission in advance, you will be able to stop only on the southernmost island along with the rest of the general public, and the odds are three in four that the most interesting birds will be on or around the islands closed to you.

What will you see? Much depends on the season and on the force and direction of the winds, which are best for sea birds when strong and easterly. In spring and fall hope for shearwaters, jaegers, and storm-petrels, and expect cormorants, shorebirds, gulls, and terns, plus a variety of small land-bird migrants (sometimes including remarkable vagrants) in the grassy areas on the roadway. In waterfowl season you are almost certain to see Greater Scaup, Oldsquaw, Red-breasted Mergansers, and all the scoters. Great Cormorants and Lesser Black-backed Gulls arrive in October. In the winter months you may find Harlequin Ducks and Common and King eiders, mostly from December to February. Look among the flocks of gulls in that season for Thayer's, Iceland, and Glaucous gulls; if Bonaparte's Gulls are present, a Little or Common Black-headed Gull may be among them. Ruddy Turnstones and Purple Sandpipers are inevitable among the rocks below you, and gannets may be fishing nearby.

South of
Chesapeake Bay

The southeast corner of Virginia is known to Washington birders mainly for a southern specialty, Swainson's Warbler. Less familiar are the numerous outstanding birding areas, all within a few miles of each other, that make a weekend around Norfolk exceptionally fruitful in most months of the year.

Hog Island, a state game refuge, lies on the south bank of the James River, not far from Williamsburg. From I-64 take the Colonial Parkway to Jamestown, and cross the river on the ferry. Continue on Va. 31 4 miles to Va. 10 and turn left (east). In 7 miles turn left on Road 650 and drive 6.5 miles to the refuge entrance, keeping left at all forks and passing the nuclear plant on the right. The gate opens at 8 A.M.

This peninsula jutting into the James River has shallow impoundments, cultivated fields, and stands of loblolly pines. It attracts large numbers of Canada Geese, dabbling ducks, Hooded and Common mergansers, and Ring-necked Ducks in winter, and shorebirds and terns in fall. Brown-headed Nuthatches are resident. Rarities have included White Pelican, Wood Stork, Greater White-fronted Goose, and Fulvous Whistling-Duck.

The **Great Dismal Swamp,** southeast of Suffolk, deserves a visit from late April to mid-May, after the Swainson's Warblers have arrived to breed and while the migrants are still moving through.

This famous swamp, surveyed by George Washington, is now a National Wildlife Refuge. The easiest way to reach refuge headquarters is to go south from Suffolk on Va. 32 for 4.5 miles beyond the point where it diverges from U.S. 13. Turn left (east) on Road 675 and follow the signs. The refuge staff recommends that visitors drive north on Road 604 to 642 (Whitemarsh Road) and take the latter route to Washington Ditch. Birders may want to work their way north toward the northwest corner of the refuge. They are requested not to use tape recorders, which seem to be having a negative effect on Swainson's Warbler behavior and interfering with field studies of the species.

In addition to Swainson's Warbler, breeding species include Woodcock; Chuck-will's-widow; Whip-poor-will; Ruby-throated Hummingbird; Acadian Flycatcher; Fish Crow; Brown-headed Nuthatch; Yellow-throated Vireo; Black-and-White, Prothonotary, Worm-eating, Yellow-throated, Hooded, and Kentucky warblers, and Yellow-breasted Chat; Summer Tanager; and Blue Grosbeak. Of particular interest is the southern race of Black-throated Green Warbler, "Wayne's" Warbler, which arrives to nest in early April.

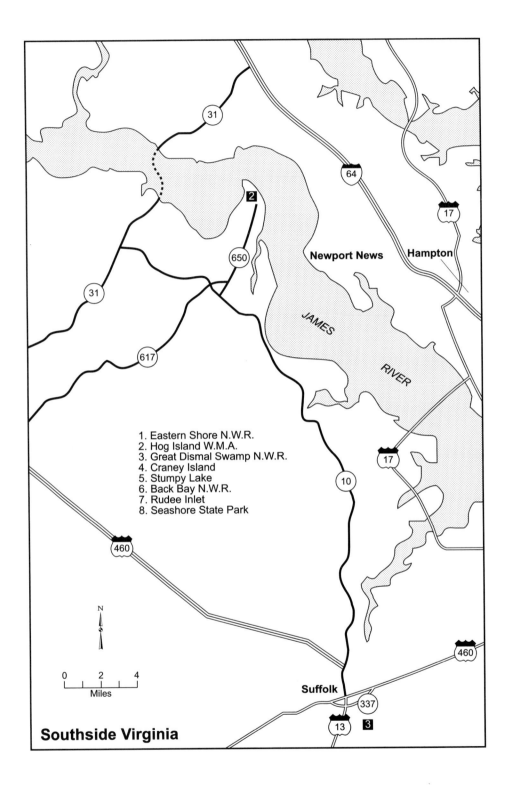

1. Eastern Shore N.W.R.
2. Hog Island W.M.A.
3. Great Dismal Swamp N.W.R.
4. Craney Island
5. Stumpy Lake
6. Back Bay N.W.R.
7. Rudee Inlet
8. Seashore State Park

Newport News   Hampton

JAMES

RIVER

Southside Virginia

0   2   4
Miles

N

Suffolk

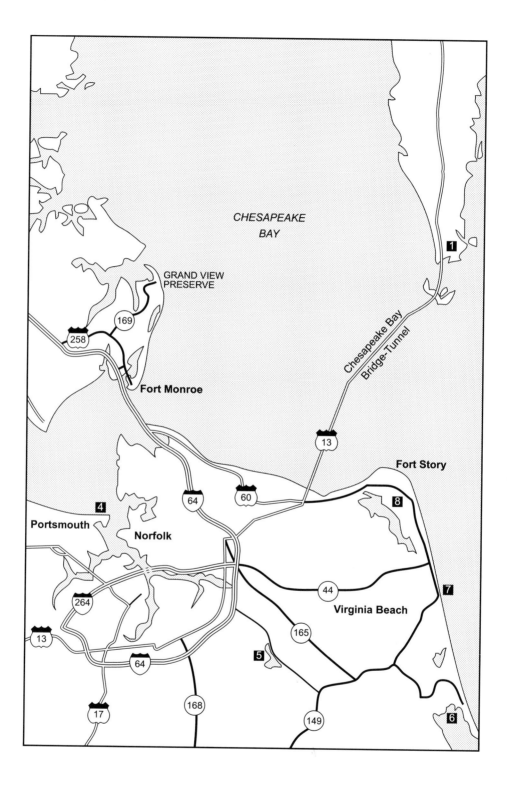

CHESAPEAKE
BAY

GRAND VIEW
PRESERVE

169

258

Fort Monroe

Chesapeake Bay
Bridge-Tunnel

1

13

Fort Story

8

64

60

4

Portsmouth

Norfolk

264

44

Virginia Beach

7

165

13

5

64

17

168

149

6

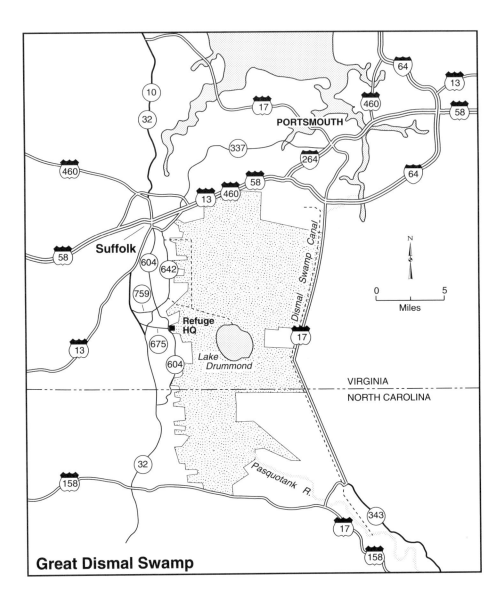

Great Dismal Swamp

Craney Island Landfill, often a superb area for water birds, is open weekdays only. From the junction of U.S. 17 and I-264 in the city of Portsmouth, head northwest on U.S. 17 for about 3.5 miles until you see a sign for Churchland Park. Turn right at that point on Cedar Lane and follow the signs to Craney Island Landfill, turning left after 1.9 miles on River Shore Drive and then right after 1.0 mile on Hedgerow Lane, which will take you to the gate. Inside, fork left at the Y-junction and go to the U. S. Army Corps of Engineers office, where you must register, at the end of the road. Then go back to the first left, and drive around the enormous impoundment formed from the spoil dredged from the great harbor beyond. You may go anywhere your car or feet will take you provided that you keep out of the way of the work trucks. A few tongues of dikes stick out into the impoundment, most of which are worth exploring, and sometimes you will want just to run up onto the peripheral dike to check what may be below.

What you should find, in their season, are hundreds, maybe thousands, of waders, waterfowl, shorebirds, gulls, terns, and skimmers. Along the west side of the impoundment are endless tidal flats and a cove where a Western Grebe lingered one winter. Look for phalaropes and godwits, Ruffs and Curlew Sandpipers, or Snow Buntings and Lapland Longspurs if you do not get there until November. Offshore in the colder months you may see Horned Grebes, sometimes Eared or Red-necked grebes, Shovelers, Gadwall, Canvasbacks, Wigeon, Ruddy Ducks, and Red-breasted Mergansers.

As in all impoundments in these parts, water levels are dependent on rainfall, and the spots where the birds are to be found cannot be predicted. Explore for yourself. Be careful not to disturb the nesting species.

Stumpy Lake. From I-64, take Indian River Road east about 3.4 miles, turn right at the sign for Stumpy Lake Golf Course, and drive to the end. Park only by the clubhouse. The swampy, man-made lake is bordered by pines and deciduous woods and provides good birding all year.

Red-headed Woodpeckers, Brown-headed Nuthatches, and Pine Warblers are resident, and Yellow-throated Warbler is a nesting species. Keep to the edges of the golf course and try not to distract the golfers. Common and Hooded mergansers and Ring-necked Ducks winter in the lake, Rusty Blackbirds and sometimes Orange-crowned Warblers, in the adjacent woods. In late summer look for herons, terns (especially Caspian), and shorebirds. Anhingas have frequently been present in recent years. The best place to look for them is from the dam at the far end of the lake; check the water and all the snags to the east.

Back Bay National Wildlife Refuge. From I-64 take Va. 44, the Virginia Beach Expressway, east to the first exit. At the foot of the ramp, turn left onto Newtown Road south, and left again in 0.2 mile onto Va. 165, Princess Anne Road. At the T-junction in Princess Anne, turn left on Va. 149 (mile 0.0) and follow the signs to the refuge. Bear right at the fork at 1.9 miles; turn left on Sandbridge Road at 2.6 miles and left at the T-junction at 4.8 miles. When you reach the community of Sandbridge, turn right on Sandpiper Road (at 8.1 miles) and drive to the refuge (entrance at 12.4 miles, parking lot at 13.6 miles).

The refuge is on a bay that can be scoped from a point close to the parking lot, but most of it is accessible only to the hiker. A loop road of about 7 miles encircles the three impoundments, and a spur road at the south end leads down to False Cape State Park. The west side of the loop is the better surface to walk on. To reach the beach, cross the dunes only on the road running east from the parking lot.

Back Bay has essentially the same species as Chincoteague, but the landscape and its relative isolation make a visit an entirely different experience. A sea watch here can be rewarding.

On the east side of Back Bay, along the back roads south of Princess Anne and around Pungo, Lark Sparrows and Brewer's Blackbirds have been found in winter in recent years. The blackbirds are most often seen in cattle feedlots. Birders are not particularly welcome in this neighborhood, so be punctilious about staying off private property and being especially courteous and responsive to requests that you move on down the road.

Rudee Inlet, at the south end of the ocean front of Virginia Beach, where U.S. 60 and Va. 149 meet, is a narrow channel lined by rock jetties. It is accessible by car on both sides and is worth scanning for Purple Sandpipers, Common and King eiders, Harlequin Ducks, and the more common jetty species. (See the discussion of Ocean City Inlet, Chapter 27, for possibilities.)

Seashore State Park is north of Virginia Beach and just inland from Fort Story at the entrance to Chesapeake Bay. From U.S. 13 take U.S. 60 about 4.5 miles east to the park entrance on the right.

Access is free between the Labor Day and Memorial Day weekends. Largely preserved as a natural area, this park includes habitats ranging from lake shore through swamp, pine and deciduous woods, and dune thickets to open beach and Chesapeake Bay. Among the 27 miles of footpaths, Bald Cypress, Long Creek, and Main trails offer perhaps the greatest variety. Brown-headed Nuthatches are resident, but the time to visit is in migration, from early April to mid-May and from early September to late October. Excellent in spring, it is superlative in fall for thrushes, vireos, warblers, and tanagers. Be sure to check the beach for shorebirds, gulls, and terns and the dune scrub and thickets and the marshes for sparrows and other passerines.

One of the favorite gull-watching sites in Virginia is just east of the park. The entrance to **Fort Story** lies 0.4 mile farther east along U.S. 60. Go left on Va. 305N and drive 2.3 miles. Civilians are normally admitted without difficulty. Park on the left in the parking lot for the Cape Henry Memorial and walk out to the beach. There are several vantage points that should provide you with good views of the gulls that are present. Lesser Black-backs are often present with the common species; one of the white-winged gulls is sometimes there, and a Little Gull is reported with Bonaparte's Gulls in winter more often than anywhere else in the region.

On the north side of Hampton Roads two stops are especially worth a visit. **Fort Monroe** is reached from I-64 at the north end of the Hampton Roads Bridge-Tunnel, via Va. 169N at Exit 69 (mile 0.0). Go right on Va. 143 and keep left at the guardhouse. Take the first left onto Eustis Lane, which circles lake-like Mill Creek. At 1.7 miles park in the small lot on the right and walk across to the concrete ramp on the left. The shoreline is worth checking for shorebirds in migration and the creek has a variety of waterfowl in winter. At Fenwick Road turn left and at 2.8 miles turn right and park by the seawall. Walk up on the wall to scan Chesapeake Bay. It is a particularly good spot to look for Brant, as well as sea and bay ducks. Then continue to the end of the road at 3.9 miles. Retracing your route, stop at any of the parking areas on the right and walk out to Mill Creek. Explore the various grassy islands and look along the banks for rails. The open areas are good for Horned Larks and for Short-eared Owls in winter, and you may find Horned Grebes in the creek and the bay. (While you are in Fort Monroe, be careful to obey all the traffic signs and to wear your seatbelt.)

Grandview Preserve is an especially rewarding place to see a variety of water birds despite its year-round popularity with human beings. Leave the fort on U.S. 258 (Mercury Boulevard) and drive 4 miles or so to Fox Hill Road, Va. 169. Where Va. 169 turns right, keep straight on Beach Road. At the far end of the last houses on the left, turn left on State Park Road and park near the end of the road. Walk around the barrier and take the quarter-mile trail to the beach. The ponds on the way may shelter American Bitterns, Black-crowned Night-Herons, Pied-billed Grebes, and dabbling ducks. Walk left up the beach at least the quarter-mile to the rockpile where a lighthouse once stood—and as far beyond as you like. A colony of Least Terns may still be present about 3 miles up. If you go in shorebird season look for oystercatchers, plovers, and a variety of sandpipers. In the colder months you should see loons, grebes, Brant, and sea ducks.

The North Carolina Outer Banks 33

Of course there is no need to go nearly so far afield as North Carolina for wonderful birding. The birds and the landscape of the Outer Banks barrier islands are, however, just different enough and just close enough to beckon many Washington birders that far south at least once every year or two.

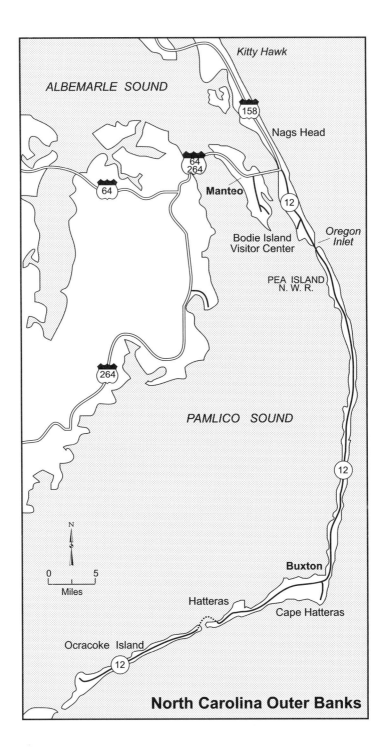

North Carolina Outer Banks

ALBEMARLE SOUND

Kitty Hawk

158

Nags Head

64
264

64

Manteo

12

Bodie Island
Visitor Center

Oregon
Inlet

PEA ISLAND
N. W. R.

264

PAMLICO SOUND

N

0   5
Miles

12

Buxton

Hatteras

Cape Hatteras

Ocracoke Island

12

Species that are much more common on the North Carolina coast than in Virginia and Maryland include (in season) Brown Pelican, White Ibis, and Sandwich Tern, plus longed-for pelagic species like Audubon's Shearwater, Black-capped Petrel, Band-rumped Storm-Petrel, and Bridled Tern, and occasional visitors like Fulvous Whistling-Duck, Swallow-tailed Kite, Long-billed Curlew, and Gray Kingbird.

The Banks are interesting all year and the beach crowds and surf fishermen from late May to November may fill the motels and campgrounds but never overrun the birding spots.

If you can, take a four-day weekend. It is about a six-hour drive to the top of the Banks, via I-95, I-64, Va. 168, and U.S. 158. If you are bored by interstate highways, leave I-95 at Fredericksburg on U.S. 17, which connects with I-64 at Newport News; it takes a bit longer, but it is a pretty and quiet route with little traffic north of Yorktown.

From Memorial Day to Thanksgiving you should have motel reservations. The most convenient accommodations for birders are in Nag's Head, Manteo, Buxton, Hatteras, and Ocracoke.

Measurements begin at the east end of the long bridge across Currituck Sound to the Banks, and detours from the highway are not included in the cumulative mileage. Small green signs with white numbers mark each mile from the bridge.

At 0.2 mile there is a garden supply company on the right, with Barlow Road, a dirt road, running south from U.S. 158. Turn right on Barlow and take the first right on Amadas Avenue. Wherever there is a fork in the road, follow the power lines toward the sound. This whole area is good for nesting and migrating warblers and other songbirds. Swainson's Warblers have been nesting here recently.

At 0.5 mile a traffic light marks the intersection of U.S. 158 with Dogwood Trail to the north and Main Road to the south. Kitty Hawk Elementary School is on the northeast corner of the intersection. In migration (mainly mid-April to mid-May and September-October), transient woodland passerines can be abundant both north and south of the intersection for a mile or so. (If you go north, park at the school or well off the pavement to avoid a parking ticket.) See what you can find, investigating any tracks, trails, or clearings. Swainson's Warblers have nested in the canebrakes along the stretch south of the highway; they may disappear as the land is developed.

Northwest of the school a pond parallels the road about 100 yards to the left. You cannot see it from the road, but a walk downhill and through the woods to its edge will pay off in views of herons, woodpeckers, sparrows, and surprises. In summer the woods are often quiet and in winter almost birdless.

At the traffic light at 4.3 miles (with a 7-Eleven store on the southeast corner), turn right on Kitty Hawk Road. Go 0.9 mile and turn left onto Herbert Perry Road just past the Unitarian Universalist church. Follow this road 0.3 mile and search the roadside ditches from here to the end of the pavement. This area is particularly good in winter for rails and bitterns. (Black Rails have been seen here.)

Return to U.S. 158 and continue to the entrance to Wright Brothers National Monument at 7.5 miles. The monument is worth a visit for itself alone. (A fee is charged from Memorial Day to Labor Day.) In addition, the well-mowed grass, especially after a hard rain, sometimes attracts Lesser Golden-Plovers and Upland, Pectoral, Baird's, and Buff-breasted sandpipers in fall migration and Horned Larks and American Pipits in winter.

At 16.0 miles, the highway becomes N.C. 12 as it enters the Cape Hatteras National Seashore at the intersection with U.S. 64 and U.S. 264.

Beginning at 18.9 miles, a series of shallow ponds line the west side of the road. The first, often the most productive of interesting birds, is set well back and is easy to overshoot. Sedge Wrens and wintering Common Snipe live in the marsh grass; ducks and gulls linger in winter, and shorebirds are abundant in spring and fall if the water levels are low.

Two of the larger ponds have small observation platforms and paved parking pull-outs. Check them all if you have time and the light is right.

At 21.9 miles, a left turn leads to Coquina Beach, a popular picnic area and bathing beach and the most accessible one on Bodie Island. A sea watch here can be rewarding from October to April. A right turn from the highway takes you to Bodie Island Visitor Center with its lighthouse and pond. The mile-long entrance road to the visitor center, lined with pines, is very attractive to land-bird migrants, but the glory of Bodie Island (which has not been an island in living memory) is its huge pond with marshy edges and islands. It is overlooked by two observation platforms, one 8 to 10 feet above the marsh and the other at the water's edge.

Depending on season and water level, the pond is notable for its Tundra Swans, dabbling ducks, herons and egrets, freshwater rails, shorebirds, and gulls, and a favorite hunting ground for falcons and accipiters. Marsh and Sedge wrens, Swamp Sparrows, and Common Yellowthroats haunt the edges; Rufous-sided Towhees and Grey Catbirds mew in the shrubbery; and Brown Creepers have been known to spiral up the lighthouse. Visit on a cloudy day or in the afternoon for the best light.

At 24.0 miles, detour into the Oregon Inlet Marina and drive along its east side to the end of the road. It is a good spot to check for loons, grebes, and bay ducks in winter, and shorebirds, terns, skimmers, and Brown Pelicans in the warmer months. If you have a four-wheel-drive vehicle, take the sand road to the beach, a left turn just south of the road to the marina. It is a good spot for loafing gulls, feeding shorebirds, and a close view of the inlet.

Continue over the Oregon Inlet bridge onto Hatteras Island and into Pea Island National Wildlife Refuge. In winter, park just south of the bridge and walk out onto the catwalks along the bridge. Look below you on the rocks, the jetty, and the bridge buttresses for Harlequin Ducks and Purple Sandpipers.

The next road to the left angles back sharply to the old Coast Guard station. Park along the shoulder and walk around the barrier at the end of the road. The stand of wind-blown trees ahead on the left is a fine migrant trap, especially in autumn. The long breakwater on the south side of the inlet is another excellent place to look for wintering sea ducks, gulls, and rock-loving sandpipers.

Down the highway again, at 29.9 miles pull off into the eroding little parking lot on the right and take your scope out to the observation platform that overlooks excellent shorebird flats on the right (the Salt Flats) and the north end of North Pond on the left. The northern tips of the long islands in North Pond are a heron roost in summer and this dike is an exciting place to end the day. Shorebirding is excellent at times along the east side of the pond. In winter, swans, Snow Geese, and both dabbling and diving ducks are common from end to end. The energetic birder may find the 4.5-mile walk around the periphery worthwhile—the islands hide many birds from the non-walker's view.

At the southeast corner of the pond there is another parking lot, along with rest rooms and an observation platform with fine views of the south end of North Pond and the north end of the New Fields impoundment. The former area has nesting Common Moorhens and, if mud flats and sandbars are exposed, a good variety of shorebirds and terns. The New Fields are superb for shorebirds in summer (especially the more long-legged species like American Avocet and Black-necked Stilt), attractive to ibis and herons, and full of geese and dabbling ducks in winter. Between the impoundment and the road, marshy fields are planted for Canada and Snow geese, occasionally joined by a Ross's Goose and more rarely by a Greater White-fronted Goose. This is the best place in North Carolina for Ring-necked Pheasant.

Across from the rest rooms a trail leads to a lookout point high in the dunes, a prime spot for an afternoon sea watch. In summer you may not see much but shorebirds and terns, but in winter you can almost count on loons and gannets, scoters and mergansers, and an assortment of gulls. With patience and luck you may see a Pomarine or Parasitic jaeger, a Great Cormorant (on the remains of a shipwreck below you), or even a Black-legged Kittiwake far offshore.

The New Fields impoundment (except for views of it from the North Pond dike) and the South Pond farther down the road have recently been closed to independent birders. You must be escorted into these areas by a volunteer with whom you have made arrangements in advance. Call (919) 987-2394 to get more information. Otherwise you can bird these ponds, which are rather distant, only from the road. Parking on the shoulder gets really risky toward the south end of New Fields; there are a few firm spots on the east side of the highway a little beyond the dike that divides New Fields from South Pond. At 32.5 miles, a sand road (closed to visitors) runs off to the west around the north end of South Pond. At refuge headquarters at 33.8 miles, a refuge leaflet and map and a bird list are available and a sighting list is displayed. Make a habit of checking the list every time you pass headquarters and of entering notable birds you have observed.

Most birders drive nonstop from this point to the turnoff to Cape Hatteras Lighthouse at 62.8 miles, a high-priority stop even when the birding is slow.

From the entrance, drive past the sometimes interesting pond on the left to the intersection at 0.9 mile. A left turn takes you to the visitor center (rest rooms, good exhibits, interesting bookstore) and the lighthouse, which is closed to visitors. You may want to scan the ocean here.

A right turn at the intersection takes you past the start of a nature trail through a maritime forest to a Y-junction. The right-hand road leads to a National Park Service campground; the left-hand road leads to a small pond best viewed from the hood of your car and to the jeep trail out to the beach. If you do not have a four-wheel-drive vehicle, walk, at least as far as the huge salt pool that is conspicuous to the right.

The entire beach around the point is so deeply rutted by fishermen's vehicles that walking is very unpleasant. If you can endure it, though, a hike out to the very point can pay off, at least from October to May, in spectacular counts of seabirds, including shearwaters, gannets, and jaegers, while shorebirds may thread their way through the fishing poles. During the last half of May, you might witness a good movement of shearwaters, jaegers, and other pelagic species, especially during and after east winds.

The concentration of gulls, terns, skimmers, ducks, and shorebirds around the tidal pool almost always yields a rarity or two. Check the tern colonies nearby in summer for an occasional Roseate or Sooty Tern. Piping Plover is often seen here except in the coldest months. In winter a few Lesser Black-backed Gulls are often present. Check the Bonaparte's Gulls, especially from February to April, for a Little Gull. A few Lapland Longspurs and Snow Buntings are seen in the low dunes every year.

(If you have a four-wheel-drive vehicle, your perspective will change completely and you can have the time of your life birding the beaches both here and farther south, around the tip of Hatteras Island and on the beaches of Ocracoke Island.)

Drive on from here down the highway to a National Park Service parking area on the right at 70.2 miles. A boardwalk takes you out to a deck overlooking a salt marsh frequented by herons and sometimes shorebirds. You will reach the village of Hatteras at 71.2 miles. Check the pilings in the harbors on the north side of the highway for cormorants, pelicans, gulls, and terns.

The road ends at 73.9 miles, at the slip for the free ferry to Ocracoke Island. If you have an extra day, spend it on Ocracoke (especially if you have four-wheel drive). It is an island small enough to explore on your own. Land birding can be outstanding around the village in September and October. Black Rails have been found in the marshes on the island, particularly those along a sand track on the left just before you come to the village. A nature trail across from Ocracoke Campground goes through woods and leads to an observation platform over the salt marsh.

If you have only half a day to spare, take a round trip on the ferry. It is wonderful, with birds all the way, flying around the boat, swimming in the sound, or resting on the numerous spoil islands along the way. Brown Pelicans are a certainty but anything may turn up, from a Wilson's Storm-Petrel or an eider to a jaeger or a Black-legged Kittiwake. An American White Pelican is sometimes present. If you have time, take an hour on Ocracoke to bird the newly created spoil ponds on the right just beyond the ferry slip, the older pond 300 yards up the road on the left, and the flats and marshes off the track to the right, just opposite the pond.

A single note of warning: Unless you have a vehicle with four-wheel drive, exercise extreme caution in driving any unpaved road or parking along road shoulders. Bare or sparsely covered sand is lethal to ordinary cars, and grassy stretches can be treacherous. Timid imitation of the clearly successful parking behavior of other two-wheel-drive cars will minimize your risks.

Special Pursuits

Pelagic Trips 34

The single reason that birders from all over the continent come regularly to this region is the opportunity to take boat trips out into the Atlantic Ocean. These trips, geared exclusively to enable participants to observe seabirds and marine mammals, offer the chance to see species that can rarely, if ever, be seen from land away from their breeding grounds.

Trips out of Ocean City, Maryland, are typically scheduled at least six times a year. Conditions permitting, the ninety-foot boat goes 50 to 100 miles out to the canyons at the edge of the continental shelf, where upwelling nutrients tend to attract feeding flocks of birds and mammals.

Winter species seen regularly are Northern Gannet and Black-legged Kittiwake; other possibilities include Fulmar; Manx Shearwater; Great Skua; Glaucous, Iceland, and Thayer's gulls; and Atlantic Puffin, Razorbill, and Dovekie. This is the least promising season for sea mammals.

In spring and early summer the winter species gradually disappear and are replaced by Greater, Sooty, Manx, and Cory's shearwaters; Wilson's Storm-Petrel; Red-necked and Red phalaropes; all three jaegers; and, sometimes in June, South Polar Skua. Sea mammals in this period may include Fin, Sperm, Minke, and Right whales; Pilot Whales; and Common, Bottle-nosed, Risso's, and White-sided dolphins.

Late summer and early fall typically produce Cory's and Greater shearwaters, Wilson's Storm-Petrel, and the phalaropes and jaegers. Rarities that may turn up include Audubon's Shearwater, White-faced and Band-rumped storm-petrels, Black-capped Petrel, White-tailed Tropicbird, Sabine's Gull, and Bridled and Sooty terns. Fin and Minke whales, Pilot Whales, and Bottle-nosed, Risso's Spotted, and Striped dolphins are all possibilities.

By late fall most of the warm-weather migrants have departed, although some shearwaters and jaegers are often still around; but several of the winter species can be expected, including gannets and kittiwakes.

Trips out of Virginia Beach, Virginia, are currently scheduled six or seven times a year, with at least one in February. The trips typically go out to Norfolk Canyon and stay within Virginia waters. The winter trip is in a seventy-five-foot boat, the others in a smaller, faster one (fifty-eight feet). The species to be expected are essentially the same as those listed for Ocean City. An effort will be made to avoid conflicting dates for trips out of the two ports, thereby increasing opportunities to go to sea.

At this writing two overnight trips a year, one in February and one in May, are scheduled to go out of Virginia Beach or Ocean City.

Trips also go out to the Gulf Stream from Manteo, North Carolina, three times a year: a mid-July weekend, a mid-August weekend, and Labor Day weekend.

Species that can reasonably be expected are Cory's, Greater, and Audubon's shearwaters; Wilson's Storm-Petrel; Red-necked Phalarope; and Pomarine Jaeger. There is a fifty-fifty chance of sighting Black-capped Petrel, Parasitic Jaeger, and Bridled Tern. There is perhaps a one-in-five chance of seeing Masked Booby, White-tailed Tropicbird, Leach's and Band-rumped storm-petrels, Red Phalarope, and Sooty Tern. More than a dozen other species have been recorded over the years.

Though an expedition to the Outer Banks is much more of an undertaking for Washington-area birders than one to Ocean City or Virginia Beach, the birds, in compensation, are a little closer to shore and tropical species are somewhat more probable. Typically, the boat reaches the Gulf Stream, its destination, in about two hours. The sea, however, is not necessarily calmer, nor the voyage drier.

All trips from Ocean City and Manteo and the February trip from Virginia Beach are scheduled with alternative weather dates the same weekend. The ship captain makes the final decision as to whether conditions are safe for a voyage; often the weather at the dock or close to shore is deceptively benign when the sea 20 miles out is too hazardous to venture on.

Trips offshore can be enormously exciting or fairly dull. Most pelagic species are likely to be seen far from land, with only intermittent sightings during the two hours or so between the harbor and the canyons along the continental shelf. You should take a pelagic trip only if you are prepared to be cold, wet, possibly seasick, and sometimes bored. The rewards can be great enough to compensate for all the discomfort.

All these trips are run by knowledgeable and experienced leaders who provide an abundance of useful information to participants, both before and during the voyage. For the current schedules and prices, write to the following individuals:

Ocean City: Gene Scarpulla, Atlantic Seabirds, 7906-B Knollwood Road, Towson, MD 21204; *Virginia Beach:* Brian Patteson, Post Office Box 125, Amherst, VA 24521; *Manteo:* Paul G. DuMont, 750 South Dickerson Street, #313, Arlington, VA 22204; *overnight trips:* Ken Bass, 12604 Valley View Drive, Nokesville, VA 22123.

Since names and addresses may not always be the same and other operators run pelagic trips in the area, your best chance of keeping fully informed and up to date is to become a member of the American Birding Association (see Appendix A). That organization publishes an annual pelagic trip issue of its newsletter, *Winging It,* each January, listing all scheduled trips for all three coasts.

Sites for Watching Hawk Migration 35

Washington residents are exceptionally fortunate in the ready access they have to the major routes of migrating raptors. Nationally famous hawk-watch spots like Hawk Mountain and Cape May Point are only a half-day away; outposts along the easternmost Allegheny ridges and the western shore of Chesapeake Bay are much closer; and the grounds of the Washington Monument or a backyard close to the Potomac are, on occasion, excellent places to lie on one's back and stare at the sky.

Generally, any winds from the north or northwest mean good hawk-watching weather in autumn at one lookout or another, but the best conditions are found a day or two after a major cold front has pushed through, lasting as long as the wind is blowing from the same quadrant. Strong northwest winds along the coast cause migrants to keep close to the shoreline, rounding Cape May or Cape Charles and heading north on the eastern side of Delaware or Chesapeake bay until land is clearly visible to the west. Then they cross over and head south again.

Fall migration begins in August, peaking in September for Broad-wings, Bald Eagles, Ospreys, and American Kestrels and, at the end of the month, for Peregrines. In October Sharp-shinned and Cooper's hawks and Merlins reach their highest numbers, whereas the first of November is prime time to look for Red-tails, Red-shoulders, Northern Harriers, Turkey Vultures, Northern Goshawks, Golden Eagles, and Rough-legs.

For those who do not dedicate their entire autumn to raptors, perhaps the best strategy is to aim for nearby mountains in mid-September, Cape May in late September, Assateague in early October, and Hawk Mountain from late October to early November.

Spring hawk migration is largely ignored by local birders; the most intensive studies in this area have been conducted by observers at Fort Smallwood Park. Recent spring surveys from mid-March to mid-June have averaged seventy-nine birds an hour and up to thirty-two Merlins a day—the highest count in North America! Peak counts for most species occur in late April and early May, and the best conditions are provided by winds with a westerly component.

In general, the highest counts in the day come between 9 A.M. and 11 A.M., with a secondary peak between 2 P.M. and 3:30 P.M. This schedule applies consistently across the region from the mountains to the coast.

Hawk watching is a sociable sport in which more viewers mean more birds seen by all. The lulls between birds are brightened by tips on identification, arguments about field guides, tales of adventure, birding gossip, and the chance to meet birders from all over the country.

Unless you are going to one of the observation points staffed throughout the season by experienced raptor enthusiasts, it pays to round up some friends to keep you company. Hawk watching requires patience and attentive scanning, and it takes unusual dedication to sustain this kind of behavior for long by yourself.

Dress more warmly than you think necessary, take along thermoses of hot drinks, and never forget a thick cushion for mountain rocks.

Listed below are ten favorite spots with Washington area birders.

HALF-DAY OUTINGS (one hour or less from the Capital Beltway, I-95/495)

1. **Fort Smallwood Park,** Maryland, at the mouth of Baltimore Harbor. Take the Baltimore-Washington Parkway north 16 miles from the beltway to Md. 176. Go east 5 miles (toward Glen Burnie). At Md. 3 go south 0.6 mile to Md. 100, then east 8.2 miles to Md. 607, Magothy Bridge Road, which becomes Hog Neck Road. Follow Md. 607 north 1.1 miles to Md. 173, Fort Smallwood Road. Turn right and drive 3 miles to the park entrance. According to present regulations, the park opens at 9 A.M. from November 15 to April 14, when admission is free. From Memorial Day weekend to Labor Day it opens at 6 A.M.; the rest of the year it opens at 7 A.M. The entrance fee is $3 per car from April 15 to November 14. Inside the park, take the rightmost road, go through the closed but unlocked gate, and park. If the gate is locked, park on the roadside and walk around the gate out to the beach. Alternatively, continue north to the parking area by the old fort.

   The best hawk watching is from the road along the east side of the park between these two parking areas, especially near a small building halfway between them, where raptors can often be seen skimming the treetops on the far side of the pond. In autumn hawks come from North Point across the mouth of the Patapsco River on west to northwest winds. This site is especially good in autumn for Kestrels, Sharp-shins, Ospreys, Northern Harriers, Red-tails, some Cooper's, and occasional Merlins. It is usually staffed when the wind is right. In spring, when the winds are from the south to the southwest, the viewing is often even better and notable for all species, but especially Merlins, Broadwings, and Cooper's.

2. **Snicker's Gap,** Virginia. Take Va. 7 for 41 miles from the beltway, just beyond Bluemont and 17 miles west from Leesburg. Park at the crest of the gap, in the commuters' parking lot on the south side of the highway, and watch from there. Pick a day with gentle winds; otherwise the raptors fly too high to enjoy. Totals per hour (about fourteen) so far are not impressive, but the site is newly discovered, easy to reach, and requires no hiking.

3. **Washington Monument Knob,** Maryland. Take I-270 north from the beltway 31 miles, I-70 west 4 miles to U.S. 40 Alternate (note the sign to Washington Monument State Park), and U.S. 40A 10 miles to the top of South Mountain, and turn right opposite the Old South Mountain Inn. Drive 1 mile to the parking lot and walk 1,100 feet (uphill) to the monument. The best view is from the top of the monument (thirty-five steps), where seating is limited; the best winds are west to north. All species, but not many falcons, may be seen. The site is usually staffed.

FULL-DAY TRIPS (less than three hours from the Capital Beltway)

4. **Waggoner's Gap,** Pennsylvania. Take I-270 north from the beltway 29 miles to U.S. 15, U.S. 15 northeast 59 miles to Pa. 74 at Dillsburg, and Pa. 74 west 22 miles to Waggoner's Gap. Park on the left side of the road or in the lot up the steep rough road on the right.

   If you are surefooted and wearing shoes with good traction, take the short trail from the parking lot left of the one to the power station, bear right, and climb up the rock pile to the highest point. Early arrivals get the most comfortable rocks. Footing is secure, seating flatter, but watching less fruitful in the parking lot. The species to be seen are basically those at The Pulpit, with the highest Golden Eagle Counts in the East; views are best west and north. Ravens are always present. The site is normally staffed.

5. **Rockfish Gap,** Virginia. This is the best hawk-watch site in the state, about two and a half hours (128 miles) from the beltway. Take I-66 for 22 miles. Exit on U.S. 29 south and drive 82 miles to the turn-off for I-64. Go west on the 4-mile access road and 19

more miles on I-64 to Exit 19, the Afton/Waynesboro exit. Almost immediately go east on U.S. 250 and turn right just before the service station onto the entrance road for the Holiday Inn. Park in the motel lot and walk around the building to the terrace on the east.

The gap is famous for huge flights in mid-September of Broad-wings, often coming from the northeast, but it is good all fall. It is usually staffed.

6. **Point Lookout,** Maryland (see Chapter 21). Take Md. 5 south-east from the beltway to the end of the road, about 70 miles. Look for the same species as at Fort Smallwood on the same winds, both fall and spring. The site is not regularly staffed.

7. **The Pulpit,** Pennsylvania. Take I-270 north from the beltway 31 miles, I-70 west 29 miles, I-81 northeast 15 miles, Pa. 16 west 13 miles, Pa. 75 north 6 miles, and U.S. 30 west 6 miles to the crest of Tuscarora Mountain. (This is the scenic route. You can continue on I-81 to U.S. 30 and go west through Chambersburg, which is a congested city strung out along the highway.) Park on the right (north) side of the highway and cross to the south. The area is owned by a hang-glider club, and you must walk up the road that the members drive to their parking lot. At the power-line crossing, you will see the sign for The Pulpit. A trail leads to the observation point, staffed by experts from mid-August to Thanks-giving and from late March through April.

The pulpit is excellent in autumn for most species, including Golden Eagles and Goshawks, in west, north, and especially southeast winds. Numbers in spring migration are much lower.

WEEKEND TRIPS

7. **Hawk Mountain,** Pennsylvania, is about four hours from Washington via I-95, I-695, I-83, I-81 north, and I-78 east. Exit at Hamburg on Pa. 61, go north 7 miles, and turn right on Pa. 895. After 2.5 miles, turn right at the sign for Hawk Mountain Sanctuary.

This site is nationally famous and fantastically popular, with superb views. Try to be there for the two or three days of northwest winds after a cold front. Not many falcons or Cooper's Hawks pass by, but Hawk Mountain is unsurpassed for most other species of raptors in fall migration. There are two lookouts, one close to the parking area, one a rocky 0.7 mile uphill; they are always staffed with lots of experts. At headquarters there are instructive displays, a fine bookshop, and weekend evening programs in autumn. It is worth a call in advance to check on the weather and current raptor movements: (215) 756-3431.

8. **Bake Oven Knob,** Pennsylvania, is 15 miles northeast of Hawk Mountain, with the same birds, fewer people, and easier walking. Stay on I-78 for 11 more miles. Go north on Pa. 143 and turn left on 309. After 2.0 miles go right on Road 39056 for 2.1 miles. Go left on an unmarked road next to a white house and, when it bends right, keep straight instead on unpaved Bake Oven Road. Follow it up a steep hill to the Bake Oven Knob parking lots (no sign). There are two lookouts, usually staffed, north (right) along the Appalachian Trail; they offer spectacular views.

9. **Cape May Point,** New Jersey. Cape May Point is about four hours away; it takes less time but more driving via I-95 and the Delaware Memorial Bridge (choose your own route down from there). The ferry to Cape May from Lewes, Delaware (pronounced Lewis), ties you to its schedule, but provides a bonus semi-pelagic trip. From the end of the Garden State Parkway (U.S. 9), drive south on N.J. 109 through Cape May. Bear right on Perry Street, which becomes Sunset Boulevard, for 2.2 miles. Go left at the stone pillars, immediately left again, right onto Lighthouse Avenue and, after 0.7 mile, left into the parking lot. The hawk observation point is straight ahead.

This is *the* concentration point for Merlins, Cooper's Hawks, and Peregrines, with a frequent skyful of American Kestrels and Sharpshins, good numbers of buteos, Ospreys, Northern Harriers, and a few eagles of both species—360° visibility makes a deck chair ideal. The site is always staffed. Recorded information, with the tape changed Mondays and Thursdays at 6:00 P.M., is available from the Cape May Bird Observatory at (609) 884-2626.

10. **Assateague Island,** Maryland and Virginia (see Chapters 29 and 30). Just west of Ocean City, Maryland, take Md. 611 south from U.S. 50 and follow the signs to Assateague National Seashore and State Park. Park as far north as you can and walk until you have a good view from bay to ocean. Alternatively, drive south from Salisbury, Maryland, on U.S. 13 to Va. 175 and go east through the village of Chincoteague to Chincoteague National Wildlife Refuge. Snow Goose Pool and the Wash Flats are the best raptor areas.

The entire island attracts more Peregrines than anywhere else on the East Coast, especially to open areas of beach, grass, mud, or sand flats. Look for them just after coastal storms or on days with northwest winds. Other falcons, accipiters, and harriers may be common as well. There is no fixed observation point, but Peregrine banders operate from a jeep on both the north end and the Wash Flats. Sit still, watch them, and they will locate the birds for you.

36 Finding Your Own Owls

For those who do not put away their binoculars after the Christmas Bird Counts and wait for the first reports of spring migration, one of the most challenging and rewarding birding pursuits is the search for wintering owls.

Because there is very little published information on finding owls, the average birder usually feels that setting out to see any owls on one's own is a hopeless enterprise. It is not, of course, though only rails require comparable initiative. You do not need to be led to an owl by the hand, though you can often see more of them that way (with the same owl and its habitat suffering more disturbance on each visit).

The favorite owling spots around Washington and some of the resident owls have long been overvisited by local birders, who sometimes pursue their goal with total disregard for the well-being of the birds or their habitat. As a result, field trips to see woodland owls have been discontinued and publicity is no longer given to sightings of any species but wintering Short-eared and Snowy owls, which are open-country birds that can easily be viewed at a distance. Accordingly, the only locations pinpointed in this book are for birds that are either unapproachable or relatively unflappable. Some areas to search in are recommended, though.

The following suggestions are designed to help you find owls independently with a minimum of harassment and environmental destruction. (Be prepared for failure, though—owls are hard to find.)

1. Learn to recognize appropriate habitat. In particular, bone up on varieties of conifers, such as red pine, spruce, and arbor vitae, that owls like most. Parks, state forests, reservoir borders, cemeteries, old nurseries, and even small roadside plantations are all good possibilities. Be alert for such areas all year long and work through them from December to March.
2. Purchase or borrow a strong flashlight, a small portable tape recorder, and a tape with a minute or so each of calls of Eastern Screech-, Barred, Barn, and Great Horned owls. You may want to try a tape on Saw-whet and Long-eared owls, but these migratory species are usually silent on their wintering grounds. You can make a tape, rather laboriously, from the Peterson records or cassettes or the National Geographic Society record, or, if you're lucky, copy one from an experienced owling friend.

OWLING BY DAY

1. Walk slowly and quietly through appropriate habitat, looking down on the lower branches and the ground for white droppings and the dark, usually oblong pellets of fur and bone that owls cough up daily.
2. When you find a tree marked by either or both signs, look up into the tree for a dense spot that may hide or be an owl.
3. As soon as you spot an owl, back off immediately to the farthest spot from which you can see it. The owl is then more likely to relax, less likely to fly away.
4. If you're birding with others, make it a small group (the fewer the better) and avoid surrounding the owl. If it can't see all of you at once without turning its head, it will probably leave.
5. Pay heed to the loud, frenzied scolding of birds mobbing a predator. Though they may be bedeviling a hawk or a snake, their target is just as likely to be an owl. The smaller birds from chickadees to jays gang up on screech-owls; when crows join in, they are trying to drive off one of the large owls. Be alert; they often succeed.

OWLING BY NIGHT

1. Choose a calm, dry night on a quiet road by a woodlot or forest, preferably well away from an occupied house, especially one with a barking dog!
2. In the dark, play the voice of each owl a few calls at a time, with pauses in between like a real owl, so that you can hear the response.
3. Unless you are trying to attract one particular species, start with the voice of a screech-owl and gradually work up to that of a Great Horned Owl. Don't shift to a larger species until you give up on the previous one. (If you try to go from large owls to small ones, the latter will not respond because they can be preyed upon by the former.)
4. If and when you do get a response, do not continue more than five minutes longer. That should be enough to bring the owl in, and more playing is likely to be counterproductive.
5. Watch for the silhouette of the owl flying in to a branch overhead. If you see or hear it, turn your flashlight on and shine it as accurately as you can. Waving the beam around a lot may drive the bird away. (Pinning an owl down in a light beam takes talent and practice.)

Note: If you can learn to imitate owl calls passably, your need for equipment will be drastically reduced. As a dividend, screech-owl imitations by day do wonders in attracting small passerines.

ABSOLUTE DON'TS

1. Don't destroy habitat or cover; move carefully.
2. Don't break branches and twigs for a photograph or a better view.
3. Don't flush an owl deliberately; especially don't throw things at it or its nest. Trying to drive a bird off its nest is inexcusable.
4. Don't climb or shake the tree.
5. Don't handle an owl, even an irresistibly tame one.
6. Don't make unnecessary noise or overdo the use of tapes.
7. Don't reveal the location of an owl to anyone you don't know, especially to a nonbirder.
8. Don't go back often or spread the news widely, except for wintering Short-eared and Snowy owls.
9. Don't owl in large groups (i.e., more than a small carload).
10. In the case of wintering Short-eared and Snowy Owls, don't leave the road and don't harass them.

HABITATS AND HABITS

Saw-whet Owl. Winters mainly in the piedmont and coastal plain. Prefers arbor vitae, spruce, honeysuckle thickets in the woods. Sits tight on thin horizontal branches, usually near eye level unless frequently disturbed. Normally silent in winter. (To hear one calling, try Cranesville Swamp on the Maryland–West Virginia border in May and June.) Keep its presence hidden from strangers.

Eastern Screech-Owl. May be found anywhere year-round: all sizes of deciduous trees; young thick evergreens, including hemlock, spruce, arbor vitae; wood duck and owl boxes; woodpecker holes; knotholes. Best found by tapes and imitation at night, mobbing birds by day. Some are easily flushed, some very tame.

Short-eared Owl. Winters in coastal marshes and weedy fields in which it rests by day. Not easily flushed (flies at ten- to thirty-foot approach). Best seen when hunting at dusk or on dark afternoons, when it is normally indifferent to large numbers of well-behaved observers. Try Port Mahon and Broadkill marshes in Delaware; Elliott Island and Deal Island, Maryland; Chincoteague and Saxis marshes; Manassas Airport in Prince William County, Virginia.

Long-eared Owl. Roosts, often in groups of several birds, in conifers fifteen to forty feet high, usually in thick stands of red cedars, spruces, or loblolly, red, white, and Virginia pines. Sits tight in winter at ten to thirty feet close to a trunk about its own thickness. Check trees all around you every few steps, from all angles. Normally silent in winter.

Barn Owl. Resident in barns, silos, attics of abandoned buildings, nest boxes, and red cedars and white and loblolly pines. Easily flushed from trees and buildings, less so from silos. When checking a building for the presence of a Barn Owl, one observer should stay outside to watch for it, as the inside observer may unknowingly flush it. Quite responsive to tapes.

Barred Owl. Resident in both coniferous and deciduous trees, often large ones. In upland woods in winter more often than in summer. Common year-round in bottomland woods, especially along the Potomac and Patuxent rivers. Easily flushed from a considerable distance. Hoots mostly from February to April. Responsive to tapes.

Great Horned Owl. Resident in medium to very large conifers and deciduous trees, using the bulky branches to hide its size. Often high in the crown, but may roost close to the trunk like a Long-eared Owl. Most vocal in winter because of its early courtship (December, January). Nests in January and February, and abandons eggs readily if disturbed. Responsive to tapes. Crows mobbing this species may be joined by Red-shouldered Hawks.

Snowy Owl. An occasional visitor to the area. Has turned up on government office buildings, suburban roofs, haystacks. Birds that linger more than a few days are usually found near salt water on high, open perches. So conspicuous that protection from the public is impossible, but harassment should be kept to a minimum. (Most likely to be flushed by photographers with inadequate lenses.)

Appendixes

Audubon Naturalist Society of the Central Atlantic States
8940 Jones Mill Road
Chevy Chase, MD 20815
(301) 652-9188

The oldest and largest society in the area, ANS has a vigorous schedule of field trips (preponderantly bird walks); field courses and activities for adults, families, and children; and natural history tours all over the continent, many of which are focused on birds. It publishes a newspaper, *The Audubon Naturalist News* (containing one of the strongest sections of natural history book reviews in the country) ten times a year, and an annual journal, *The Atlantic Naturalist.* It is deeply involved in regional and local conservation issues. It cosponsors a lecture series with the Smithsonian Associates and the Friends of the National Zoo and a program of Natural History Field Studies with the Department of Agriculture Graduate School. It operates the Audubon Bookshop, specializing in natural history books and recordings. It sponsors the Voice of the Naturalist, a weekly telephone recording, (301) 652-1088, that combines a rare bird alert and other birding news and information.

Maryland Ornithological Society
Cylburn Mansion
4915 Greenspring Avenue
Baltimore, MD 21209

MOS publishes the state ornithological journal, *Maryland Birdlife,* and a bimonthly newsletter, *The Yellowthroat.* It holds an annual weekend of intensive field trips in May and operates wildlife sanctuaries across the state, several with accommodations for members. Of its fourteen local chapters, the Montgomery County chapter is one of the most lively, attracting active birders from all over the Washington area. There are monthly meetings from September to May and a regular schedule of field trips, a newsletter (*The Chat*), and an annual calendar of events. To get in touch with a chapter officer (who can tell you when and where the chapter meets and what field trips are coming up), write to MCC/MOS, Post Office Box 59639, Potomac, MD 20859-9639.

Virginia Society of Ornithology
c/o Mrs. J. H. Dalmas
520 Rainbow Forest Drive
Lynchburg, VA 24502
(804) 239-2730

VSO publishes the state ornithological journal, *The Raven,* and a quarterly newsletter. It conducts a week-long foray in early summer to a Virginia county to look for breeding birds, and schedules an annual meeting with field trips in May and several birding weekends a year. It is possible to belong to a local VSO chapter without being a member of the state society, and vice versa. The Northern Virginia chapter has occasional meetings and a strong schedule of local bird walks, announced in its newsletter, *The Siskin.* For current informaton write to the Northern Virginia Chapter VSO, Post Office Box 5424, Arlington, VA 22205.

Fairfax Audubon Society (Vienna, Virginia)
Prince George's Audubon Society (Bowie, Maryland)
Southern Maryland Audubon Society (Clinton and LaPlata, Maryland)

These three clubs—west, east, and south of Washington, respectively—are the nearest chapters of the National Audubon Society, which can tell you how to make contact with the chapter president or membership chairman. Call (202) 547-9009. All chapters have regular meetings and field trips. If you are a member of the national society, you are automatically a member of a local chapter if you live within its boundaries. If you do not, you may become a member of any chapter you choose by specifying that the appropriate part of your dues go to that chapter. Members receive *Audubon* magazine.

Delmarva Ornithological Society
Post Office Box 4247
Greenville, DE 19807

Washington-area birders usually join DOS to get the state ornithological journal, the *Delmarva Ornithologist,* since most society activities are in Delaware. Regular meetings (at the Delaware Museum of Natural History in Greenville) and field trips are supplemented by a spring birding weekend and long-range censuses of tracts of ornithological importance.

Raptor Society of Metropolitan Washington
Post Office Box 482
Annandale, VA 22003

Anyone interested in raptors is welcome to join this organization, which meets monthly at the National Wildlife Federation in Vienna, Virginia, publishes a newsletter, and conducts occasional field trips.

American Birding Association
Post Office Box 6599
Colorado Springs, CO 80934

ABA is the national organization for birders. It publishes *Birding,* a bimonthly magazine that focuses on solving problems of field identification; providing bird-finding information in North America and around the world; studies of rare, little-known, or declining species; translating into nontechnical language the findings of ornithological research of interest to birders; and reviewing books and journals, recordings and video tapes, optical equipment, and software for the birder market. Its monthly newsletter, *Winging It,* features alerts on threats to important bird habitats, news of recent vagrants, free birding-related (noncommercial) classified ads for members, and an annual special issue on pelagic trips off all North American coasts. ABA Sales gives members substantial discounts on optical equipment, books, checklists, recordings, software, and other products for birders. Biennial conventions and smaller conferences in between offer field trips and workshops in diverse locations in the United States and Canada.

Useful Publications   B

MAPS

The District of Columbia and all the states provide free road maps on request. In addition, the states sell county maps, which are almost indispensable for thorough exploration of local areas. To obtain a state map and a price list for county road maps, send a postcard to the appropriate address:

Delaware
  Department of Transportation
  Post Office Box 778
  Dover, DE 19901

Maryland
   State Highway Administration
   Map Distribution Section
   2323 West Joppa Road
   Brooklandville, MD 21022

Virginia
   Department of Highways and Transportation
   1221 East Broad Street
   Richmond, VA 23219

District of Columbia: Call (202) 724-4091 or write:
   Department of Transportation, Room 519
   415 Twelfth Street, NW
   Washington, DC 20004

To order a Maryland county topographic map, send a check for $3 per map (postpaid) to:

   Maryland Geological Survey
   2300 St. Paul Street
   Baltimore, MD 21213

Even more detailed, though rarely as up to date, are the topographic maps of the U.S. Geological Survey. An index and catalogue are available from:

   Book and Map Sales
   U.S. Geological Survey
   2201 Sunrise Valley Drive
   Reston, VA 22092

If you love maps, you will enjoy visiting this office, where visitors are very welcome. There is also a downtown sales office at 1951 Constitution Avenue, NW, in Washington.

   Of particular value to birders is an atlas of topographic maps for the entire state of Virginia. Though you may want a magnifying glass to see the details, it shows every back road in the state. It is available in drugstores, supermarkets, and convenience stores all over the state and in the District of Columbia. Look for the *Virginia Atlas and Gazetteer* (Freeport, Maine: DeLorme Mapping Company, 1989).

   A new map of the Potomac River and the C&O Canal is available from:

   Interstate Commission of the Potomac River Basin
   Suite 300
   6110 Executive Boulevard
   Rockville, MD 20852

REGIONAL BOOKS AND CHECKLISTS

Halle, Louis J. *Spring in Washington.* Baltimore: Johns Hopkins University Press, 1988.
A classic account of the spring migration of 1945 in and around Washington, dated only in detail. Evocative writing, enhanced by drawings by Francis Lee Jaques.

Harding, John J., and Justin J. Harding. *Birding the Delaware Valley Region.* Philadelphia: Temple University Press, 1980.
All you need to know about birding Cape May, Hawk Mountain, and some seventy other major and minor locations within a two-hour drive from Philadelphia, including Brigantine and Tinicum Marsh.

Hess, Gene K., Richard L. West, Maurice V. Barnhill, III, and Lorraine M. Fleming. *Birds of Delaware.* Pittsburgh: University of Pittsburgh Press. In press.

Kain, Teta, ed. *Virginia's Birdlife: An Annotated Checklist,* 2d ed. Virginia Avifauna Number 3. Virginia Society of Ornithology, 1987.
Accidental, extinct, and hypothetical species are incorporated into the main text. For all species not in these three categories, seasonal abundance, breeding information, and out-of-season records are provided for each of the three major physiographical provinces.

Meanley, Brooke. *Birdlife at Chincoteague and the Virginia Barrier Islands.* Centreville, Maryland: Tidewater Publishers, 1981.
————. *Birds and Marshes of the Chesapeake Bay Country.* Cambridge, Maryland: Tidewater Publishers, 1975.
————. *The Great Dismal Swamp.* Washington, D.C.: Audubon Naturalist Society, 1973.
Chatty and informative reports of field work by the author and other ornithologists.

Mountains to Marshes: The Nature Conservancy Preserves in Maryland. The Nature Conservancy Maryland Chapter, 1991.
Detailed information on twenty-seven TNC preserves and on five other natural areas in Maryland that have received TNC support. Directions, maps, and visitor guidelines for fifteen to which there is public access.

Pettingill, Olin Sewall, ed. *The Bird Watcher's America.* New York: Thomas Y. Crowell, 1965.
Chapters on the Cheat Mountains of West Virginia, the Great Dismal Swamp, and Hawk Mountain. Out of print.

Robbins, Chandler S., sr. ed. *Atlas of the Breeding Birds of Maryland and the District of Columbia.* Pittsburgh: University of Pittsburgh Press. In press.

Robbins, Chandler S., and Danny Bystrak. *Field List of the Birds of Maryland,* 2d ed. Maryland Avifauna Number 2. Maryland Ornithological Society, 1977.

A marvel of compression. Accidental, extinct, and hypothetical species are listed separately; the rest are coded for physiographical province and habitat, with bar graphs of seasonal abundance. Fifty good birding areas in Maryland are very briefly described.

JOURNALS

Of the publications below, only the first is obtained by subscription. The rest are received as a benefit of membership (see Appendix A). The regional editors of *American Birds* and the editors of the state journals welcome well-documented records of unusual sightings (see Appendix D) and reports on other matters of ornithological interest.

American Birds. National Audubon Society, 950 Third Avenue, New York, NY 10022.

Birding. American Birding Association, Post Office Box 6599, Colorado Springs, CO 80934.

The Delmarva Ornithologist. Write to "Editor" in care of the Delmarva Ornithological Society (see Appendix A).

Maryland Birdlife. Editor, Chandler S. Robbins, 7900 Brooklyn Bridge Road, Laurel, MD 20707.

The Raven. Editor, Teta Kain, Route 5, Box 950, Gloucester, VA 23061.

C   Cooperative Birding Activities

For birders who enjoy putting their skills to use, there are dozens of activities throughout the year that depend on field observers with various levels of skill and energy. They range from the mailing of a postcard about a conspicuously marked bird to recording and analyzing all sightings in a given area over several weeks, months, or years. Many of these activities develop talents and interests that bring at least as much pleasure and satisfaction as seeing new species. Here is a sample:

May Counts. On the second Saturday in May, the Maryland Ornithological Society asks everyone birding in the state of Maryland to keep careful track of all birds seen on that date in each county. The results are compiled and published in *Maryland Birdlife.* MOS county chapters make a point of deploying their members for maximum coverage, and beginners are teamed with experienced observers.

On the same date, the Audubon Naturalist Society has a similar count, rather less structured. Anyone birding anywhere in the region is invited to submit the day's tally to the count compiler. A summary is published in the society's newspaper.

Christmas Bird Counts. Dozens of counts are scheduled in this area during two weeks and three weekends in late December and early January. Volunteers are assigned to census a specific territory in the fifteen-mile-diameter circle within which each count is conducted. Beginners are welcome; participants may take part in as many counts as time and stamina allow. A fee to cover the cost of publication of the results is charged to every participant on every count. Schedules and coordinators of nearby counts are listed in the Audubon Naturalist Society newspaper, and results are published in *American Birds.* Many local bird clubs sponsor one of the counts.

Winter Bird Counts. Some chapters of the Maryland Ornithological Society sponsor a count of birds in their county on some date between January 10 and February 10, following the pattern of the MOS May Count. The establishment of other Winter Bird Counts in the state is encouraged.

Breeding Bird Surveys. Birders who can recognize the songs and calls of the breeding birds of the region may volunteer to conduct breeding bird surveys in June: twenty-five-mile routes with three-minute stops every half-mile to count every bird seen and heard at each stop. Routes are assigned by the Office of Migratory Nongame Bird Studies, Patuxent Wildlife Research Center, Laurel, MD 20708.

VSO Foray. Each year the Virginia Society of Ornithology engages in a week-long survey of breeding birds and summer visitors to one Virginia county. Participants are welcome for all or part of the period.

Breeding Bird Census and Winter Bird Population Study. Sponsored nationally by the Cornell Laboratory of Ornithology (with results published by the Association of Field Ornithologists), the censusing of four local tracts is sponsored locally by the Audubon Naturalist Society in Rock Creek Park, Glover-Archbold Park, Cabin John Island (along the Potomac), and Scotts Run Nature Preserve. Intensive surveys of a small area in a well-analyzed habitat map all the birds observed within a period of several weeks and record the density of each species. Volunteers at all levels of experience are welcome to participate on survey teams. Call the Audubon Naturalist Society at (301) 652-9188 for more information.

North American Hawk Migration Studies. Birders who can identify hawks in flight may contribute to a continent-wide survey of hawk migration by making counts at given observation points and filling out standard reporting forms for the Hawk Migration Association of North America. (Most sites in this area that are already staffed are in need of more help.) The association's current membership secretary is Seth Kellogg, 377 Loomis Street, Southwick, MA 01077.

International Shorebird Surveys. Birders who can identify shorebirds may participate in ongoing surveys, begun in 1974, that census shorebird concentration points throughout eastern North and South America. A survey of the same area every ten days (or more frequently, if you like) from April to early June and from July to October is requested by the coordinator, Manomet Bird Observatory, Manomet, MA 02345.

Reporting Marked Birds. Anyone seeing a bird painted a bizarre color or wearing a neck collar, a wing tag, colored leg bands, and so on, can perform a useful service by describing the species and its marking (as exactly as possible) and listing the precise location and the date on a postcard and sending it in to the Bird Banding Laboratory, Laurel, MD 20708. If you find a dead bird with a metal band, send the flattened band in with the same information. You will eventually get a reply telling you when and where the bird was banded and marked.

North American Nest Record Card Program. Birders can obtain cards from the Cornell Laboratory of Ornithology, 159 Sapsucker Woods Road, Ithaca, NY 14850, on which they record information on each nest they find in the breeding season: species, habitat, location, number of eggs or young, and so on. The cards are easy to fill out and add useful data to the massive files at Cornell.

Project Feederwatch. Feeder watchers record the birds at their feeders during a two-day watch period every two weeks from November to March, for a total of ten watch periods. They record the highest number of birds of each species seen at any one time. Observers are asked to watch for at least an hour at a time. A fee of $12 is charged to each participant to pay the costs of the program, including the publication of the two newsletters they receive each year. To participate, write the Cornell Laboratory of Ornithology (see the previous entry).

Constructing a Checklist. Some of the local parks have no more than preliminary checklists and most nature centers welcome help in compiling one, or correcting one that is out of date. By far the most useful records for this purpose are those taken at frequent intervals all year, with numbers of each species seen and any evidence of nesting included.

Documenting Rarities   D

The only way that a country, state, or other jurisdiction can hope to maintain an up-to-date record of the species found within its borders is to receive documentation of unusual sightings from field observers. These days birds are rarely killed to provide proof of their presence in a region: hard-to-obtain federal permits are required to collect birds or their eggs or nests. The best alternative evidence is one or more unequivocal photographs together with a detailed written description of the bird itself and of the circumstances in which it was observed. Photographs are inadequate without a written account, but a complete written description, even without photographs, can provide enough details for members of the local records committee to make a decision about accepting the record.

Descriptions of a rare bird should be written *independently* by as many observers as are present, based primarily on notes taken on the spot prior to consultation of field guides or other references. A photocopy of the original notes is valuable evidence, as are on-site sketches of key parts of the bird (or of the whole bird, of course). What is wanted is a complete verbal portrait of the individual that was seen, not a generalized field guide description of the species.

Some birds can be observed in much more detail than others, and some are easier to describe and to identify than others. The appearance of a plainly patterned bird can be described quite briefly. There is no substitute, however, for frequent practice in taking notes, and that can be done when no rarity is around. Get into the habit of carrying a small notebook and a waterproof pen in the field at all times.

Birds listed in Chapter 3 that are starred in the main text or listed among the rarities and accidentals at the end of the chapter, as well as species not listed at all, are the ones for which documentation is needed in this area.

Some of these species are not categorized as accidental in one or more of these jurisdictions (Bewick's Wren and Pine Grosbeak, for example). Nevertheless, there have been so few recent sightings that documentation is the only way to keep track of the current local status, and all the records committees will be glad to receive detailed reports of such birds.

When your documentation is as complete (and legible!) as you can make it, submit it to the appropriate records committee, with a copy to the regional editor of *American Birds:*

Delaware Records Committee
Delmarva Ornithological Society
Post Office Box 4247
Greenville, DE 19807

Maryland/DC Records Committee
c/o Erika Wilson
2032 Brooks Square Place
Falls Church, VA 22043

Virginia Records Committee
c/o Teta Kain
Route 5, Box 950
Gloucester, VA 23061

The set of guidelines that follows has been adapted from the one prepared by the Maryland/DC Committee. Completing a report in this degree of detail will ensure that you have met all requirements for any bird records committee around the world. (It does not ensure that your record will be accepted, however—acceptance depends on the quality of your observations and the perceived accuracy of your identification.)

CHECKLIST OF POINTS TO INCLUDE IN DOCUMENTATION

Some of these points will not be observed, nor are they all necessary to identify a species correctly, but omitting something you did see lessens the value of your documentation. Use this checklist to remind yourself of items that you want to include in your report.

Description of Size, Shape, Plumage

1. Size relative to other birds
2. Head: shape, colors, pattern
3. Bill: length, shape, color
4. Eye: color, ring
5. Upperparts: nape, back, rump, uppertail coverts

6. Upperwings (scapulars, coverts, flight feathers): length, colors, pattern
7. Tail: colors, shape
8. Underparts: chin, throat, breast, belly, flanks, vent, undertail coverts
9. Underwings (axillars, underwing coverts, flight feathers): colors, pattern
10. Legs and feet: length, colors, shape
11. Condition: signs of molt or wear

Number, Sex, Age

1. Number of birds
2. Age, by plumage or behavior
3. Sex, by plumage or behavior

Behavior

1. Posture
2. Vocalizations
3. Feeding method
4. Preening or sleeping
5. Flight pattern
6. Interaction with other birds

Other Birds Present

Location

1. Place name: pond, park, road, town, etc.
2. County and state
3. Habitat: vegetation, soil, water, terrain

Conditions

1. Sky: cloud cover
2. Wind: speed and direction
3. Temperature
4. Precipitation, snow cover
5. Water: depth, waves, ice
6. Duration of sighting
7. Time of day
8. Distance of bird from observer
9. Sun location relative to bird and observer

Observer Resources

1. Optics (both binocular and scope), power
2. Field guides and other literature consulted—before or after writing notes
3. Experience with this species, years birding
4. Other observers present

E Ethics for Birders

1. Put the welfare of the bird first.
   a. Do nothing that would flush a bird from its nest or keep it from its eggs or young.
   b. Avoid chasing or repeatedly flushing any bird; in particular, do not force a tired migrant or a bird in cold weather to use up energy in flight.
   c. Do not handle birds or their eggs unless you have a permit to do so.
   d. Make a special effort to avoid or stop the harassment of any bird whose presence in the area has been publicized among birders. This stricture especially applies to the use of tapes and to the disturbance of nesting birds, and of vagrants and rare, threatened, and endangered species.
   e. If you think a bird's welfare will be threatened if its presence is publicized, document it carefully and report its presence only to someone who needs to have the information (e.g., a refuge manager, an officer of the appropriate records committee, the editor of the appropriate journal). If you are not sure, discuss it with the manager of a rare bird alert or another experienced and responsible birder.
2. Protect habitat.
   a. Stay on existing roads and trails whenever possible.
   b. Leave vegetation as you find it: do not break it or remove it to get a better view, or trample marshland into mud.
3. Respect the rights of others.
   a. Do not trespass on property that may be private, whether or not "No Trespassing" signs are posted. Ask the landowner directly for access unless specific permission for birders to enter the area has been announced or published.
   b. Do not enter closed areas of public lands without permission.
   c. If you find a rare bird on land that is closed to the public, do not publicize it without describing the possible consequences

of doing so to the owner and obtaining appropriate permission.

d. Stay out of plowed or planted fields and managed turf or sod.

e. By behaving responsibly and courteously to nonbirders at all times, help to ensure that birders will be welcome everywhere. Do nothing that may have the consequence of excluding future birders from an area.

f. When seeking birding information from others call only between 9 A.M. and 9 P.M. unless you know that your call will be welcome at that number at other hours.

Index

Map page numbers are in **bold-faced** type. The main entries describing birding sites are in *italics*. National Wildlife Refuge is abbreviated as N.W.R., Wildlife (Management) Area as W.(M.)A.

A

Abbott Wetland Refuge, Jackson Miles, *71*
Accipters, 56, 59, 62, 200
Accokeek Creek, 24, 33, *91*
Accotink Bay Wildlife Refuge, *70–71*
Alapocas Woods Park, 167
Albatross, Yellow-nosed, 47
Alexandria, 63–70
Algonkian Regional Park, *102–3*
Allegany County, Md., 5, 120–26, 126
Alleghany Mountain, *119*
Allen's Lane, 171
All Faith Church, 150
American Birding Association, 227, 241
Anacostia Naval Air Station, 51, *63*
Anhinga, 12, 204, 213
Ani, Groove-billed, 47
Anna, Lake, *106–9,* **107**
Appalachian Plateau, 4, 5, 6, 9, 126–30
Assateague Island
    National Seashore, 193, 198, *231*
    North End, *192–95*
    State Park, *192–95, 231*
Assawoman State W.A., *185–86*
Audubon Bookshop, 239
Audubon Naturalist Society, xiii, 239, 245, 246
Augustine Beach, *164*
Avocet, American, 21, 63, 67, 68, 125, 199, 219

B

Back Bay N.W.R., 213–14
Backbone Mountain, 5, *129*
Back River Waste Water Treatment Plant, *135–37,* **136**
Bake Oven Knob, *231*
Ball's Bluff, 104
Baltimore County, Md., 7, **134**
Baltimore Harbor, 133, 139–43, **140**
Battery Kemble Park, *53–54*
Battle Creek Cypress Swamp Sanctuary, *152*
Baxter Tract, *164*

Beaver Branch, *164*
Beaverdam Reservoir, *105–6*
Belle Haven Marina, 68
Belle Haven Picnic Area, *68*
Big Meadows, 114, *115*
Big Pool, *121*
Bird Clubs. *See* Ornithological Societies
Birds, marked, 246
Bittern(s)
    American, 12, 52, 53, 54, 70, 121, 137, 162, 180, 186, 191, 215
    Least, 13, 68, 70, 80, 81, 93, 94, 101, 133, 137, 146, 162, 166, 190, 191
Blackbird(s)
    Brewer's, 45, 159, 172, 177, 214
    Red-winged, 45, 52, 81, 101, 172, 187, 193
    Rusty, 45, 52, 68, 70, 77, 81, 90, 92, 93, 94, 150, 213
    Yellow-headed, 45, 161, 172, 173, 177, 187
Black Hill Regional Park, **86,** *88*
Black Marsh State Park, *138*
Blackwater N.W.R., 173, **174,** 179
Blairs Valley, *120–21*
Blairs Valley Lake, 121
Blockhouse Point, *79,* 82
Bluebird
    Eastern, 34, 53, 76, 77, 80, 81, 89, 92, 93, 103, 114, 118, 121, 133, 150, 175
    Mountain, 48
Bluegrass, *118*
Bobolink, 45, 55, 91, 94, 100, 105, 116, 118, 119, 126, 129, 157, 166, 187
Bobwhite, Common (Northern), 19, 60, 70, 84, 150
Bodie Island Visitor Center, 218
Bogle's Wharf Road, 171, 172
Bombay Hook N.W.R., *157–60,* **158,** 172
Booby
    Brown, 47
    Masked, 226
Books, 243–44
Booten's Gap, 115
Brandywine Creek State Park, *166–67*

Green-backed, 13, 52, 52, 72, 77, 80, 81, 85, 92, 93, 101, 118, 125, 189, 204
  Little Blue night-, 13, 51, 66, 204
  Tricolored, 13, 67, 204
  Yellow-crowned Night-, 13, 52, 53, 58, 70, 80, 105, 137, 199, 202
  See also Egret; Ibis
Herrington Manor State Park, *127–28*
Hickey Hill, 59
Highland County, *116–20*, **117**
High Winds Road, **195**
Hog Island State Game Refuge, *209*
Hooper Island, 173, **174**, *176*
Hughes Hollow, 7, *80–82*
Hughes Road, 82
Hummingbird
  Ruby-throated, 29, 70, 78, 80, 94, 121, 150, 209
  Rufous, 47
Hungerford flats, 195
Hunting Bay, *68*
Huntley Meadows Park, *70*
Hunting Quarter Road, 80, 81, 82

I

Ibis
  Glossy, 14, 62, 162
  White, 14, 159, 204, 205, 217
  White-faced, 47
  See also Herons; Egrets
Indian River Inlet, 12, *186*
International Shorebird Surveys, 246

J

Jack Mountain, 6, *118*
Jaeger(s)
  Long-tailed, 24, 225
  Parasitic, 24, 66, 67, 219, 225, 226
  Pomarine, 24, 201, 219, 225, 226
Jay, Blue, 32
Jones Point, *67–68*
Journals, 244
Jug Bay Natural Area, *93–94*
Junco, Dark-eyed, 44, 113, 116, 129

K

Kent County, Md., *169–73*, **170**
Kestrel, American, 18, 119, 146, 175, 227
Key Wallace Drive, *175*, 181
Killdeer, 20, 54, 125
Kingbird
  Eastern, 31, 81
  Gray, 48, 217
  Western, 31, 193, 200, 204

Kingfisher, Belted, 29, 57, 77, 78, 80, 93, 101
Kinglet(s)
  Golden-crowned, 33, 69, 76, 119, 126, 128, 128
  Ruby-crowned, 34, 69, 76
King's Park, *87*
Kite
  American Swallow-tailed, 47, 217
  Black-shouldered, 47, 125
  Mississippi, 47
Kittiwake, Black-legged, 26, 184, 219, 221, 225
Kitts Hummock, 163
Knot, Red, 22, 137, 162, 200, 204

L

Lark, Horned, 31, 62, 91, 93, 101, 105, 106, 126, 150, 157 173, 187, 188, 190, 215, 218
Leipsic, 157, 160, 163, 164
Lewes–Cape May Ferry, 187
Life of the Dunes Trail, 195
Life of the Forest Trail, 195
Life of the Marsh Trail, 194
Lilypons Water Gardens, 7, 99, *100–101*
Limberlost Trail, 114
Limpkin, 47
Little Bennett Regional Park, **86**, *87–88*
Little Creek W.A., 10, **158**, *160–63*
Little Falls, 53
Little Levels, 195
Little Orleans, 122
Locust Spring Picnic Ground, *119*
Locustville, *203*
Logan Lane Tract, 163
Log Inn Road, *146*
Long Branch Nature Center, *73–74*
Longspur
  Chestnut-collared, 48
  Lapland, 44, 62, 101, 102, 146, 157, 159, 173, 187, 213, 220
  Smith's, 48, 192
Loon(s)
  Common, 11, 54, 62, 91, 108, 126, 171, 184, 199
  Pacific, 47
  Red-throated, 11, 62, 171, 184
Loudoun County, Va., 7, 10, 76, *102–6*
Lucketts, *104–5*, **104**

M

Mall, the, 60
Mallard, 15, 58, 60, 175
Maps, 241–42
Marked birds. *See* Birds, marked
Martin, Purple, 31, 121, 203

Rockfish Gap, *229*
Rocky Gap State Park, *125–26*
Rocky Gorge Dam, 31
Roosevelt Island, *51–52*
Ross Drive, 57
Roth Rock fire tower, 129
Rudee Inlet, *214*
Ruff, 23, 68, 166, 198, 213

S

Saint Jerome's Neck Road, 152
Saint Mary's County, Md., *147-53,* **148, 149**
Saint Michaels, 14
Saint Paul's Church, 169
Salisbury landfill, *182*
Sanderling, 22, 66, 67, 137, 181, 200
Sandpiper(s)
    Baird's, 23, 63, 66, 125, 181, 201, 203, 218
    Buff-breasted, 23, 63, 198, 201, 203, 218
    Curlew, 23, 159, 163, 213
    Least, 22
    Pectoral, 23, 62, 67, 135, 198, 203, 218
    Purple, 23, 145, 146, 153, 184, 206, 214, 219
    Semipalmated, 22, 200, 205
    Sharp-tailed, 47, 68
    Solitary, 21, 54, 55, 70, 135
    Spotted, 21, 70, 79, 85, 135, 168
    Stilt, 23, 67, 198
    Upland, 21, 62, 101, 105, 129, 141, 163, 203, 218
    Western, 22
    White-rumped, 23, 66, 181, 191
Sandy Point State Park, 7, *143–47,* **144**
Sapsucker, Yellow-bellied, 29, 77, 82
Saxis marshes, *202–3*
Scaup
    Greater, 16, 53, 54, 60, 68, 108, 126, 137, 145, 151, 153, 162, 171, 184, 186, 206
    Lesser, 16, 53, 54, 60, 68, 108, 126, 137, 141, 145, 151, 162, 171, 186
Scoter(s)
    Black, 16, 62, 172, 184, 194, 206
    Surf, 16, 62, 172, 184, 194, 206
    White-winged, 16, 62, 172, 184, 194, 206
Scott's Run Regional Park, *75*
Seashore State Park, *214*
Seneca, *80*
Seneca Creek State Park, *85–87*
Seneca Marsh, *80*
Shantytown, 195
Shearwater(s)
    Audubon's, 12, 217, 226, 226
    Cory's, 11, 225, 226

Greater, 12, 225, 226
Manx, 12, 225
Sooty, 12, 109, 225
Shenandoah Mountain, 116
Shenandoah National Park, *113–16*
Shenandoah Valley, 6, 32
Shorebirds, 51, 62, 63, 66, 67, 68, 79, 85, 94, 100, 125, 159, 166, 213
Shorter's Wharf Road, *176*
Shoveler, Northern, 15, 60, 138, 175, 182, 213
Shrike
    Loggerhead, 35, 101, 105
    Northern, 35, 143
Sinepuxent Bay. *See* Ocean City
Siskin, Pine, 4, 46, 53, 57, 59, 60, 69, 114, 128, 188
Skimmer, Black, 27, 67, 145, 199, 204
Skua
    Great, 24, 225
    South Polar, 47, 225
Snicker's Gap, *229*
Snipe, Common, 24, 51, 53, 55, 62, 63, 90, 103, 105, 114, 137, 176, 191, 218
Solomons, *152*
Sounding Knob, *118*
Southern Maryland Audubon Society, 240
South Ocean Beach, 195
South River Falls, *114–15*
South River Picnic Area, *114–15*
Southside Virginia, *209–15,* **210, 211**
Sparrow(s)
    American Tree, 42, 58, 88, 108, 146
    Bachman's, 42
    Baird's, 48
    Black-throated, 48
    Chipping, 42, 53, 55, 60
    Clay-colored, 42, 60, 166, 193, 194, 200
    Field, 42, 53, 55, 58, 80, 88, 114, 176
    Fox, 43, 58, 68
    Grasshopper, 43, 55, 60, 77, 80, 81, 85, 93, 101, 105, 150, 159, 168, 182
    Harris's, 44
    Henslow's, 43, 126
    House, 47, 58
    Lark, 43, 193, 194, 200, 214
    Le Conte's, 48, 204
    Lincoln's, 44, 56, 90, 166, 168
    Savannah, 63
    Savannah (Ipswich), 43, 55, 118, 129, 150, 188, 196
    Seaside, 43, 63, 151, 152, 161, 179, 190, 193, 196, 199, 202
    Sharp-tailed, 43, 161, 179, 190
    Song, 43, 58, 114
    Swamp, 44, 51, 70, 72, 79, 91, 93, 121, 126, 129, 129, 137, 150, 176, 218

Three-toed, 47
Wood-Pewee
  Eastern, 30, 56, 78, 189
  Western, 47
Wren
  Bewick's, 48, 247
  Carolina, 33
  House, 33, 57, 82, 180
  Marsh, 33, 51, 68, 80, 81, 90, 91, 93,
    101, 137, 144, 161, 171, 176, 177,
    179, 190, 218
  Rock, 48
  Sedge, 33, 157, 161, 181, 190, 191,
    202, 218

Winter, 33, 54, 59, 70, 72, 82, 88, 91,
  92, 113, 116
Wright Brothers National Monument,
  218

Y

Yellowlegs,
  Greater, 21, 51, 62, 67, 70, 135, 180,
    199, 203
  Lesser, 21, 51, 62, 67, 70, 135, 180,
    199, 203
Yellowthroat, Common, 40, 68, 78, 81,
  84, 85, 90, 94, 114, 180, 218

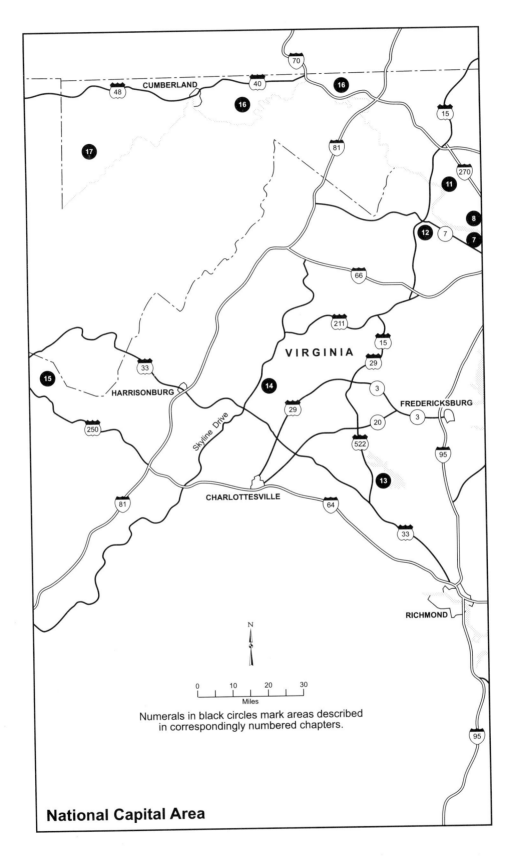

Numerals in black circles mark areas described
in corresponding numbered chapters.

National Capital Area

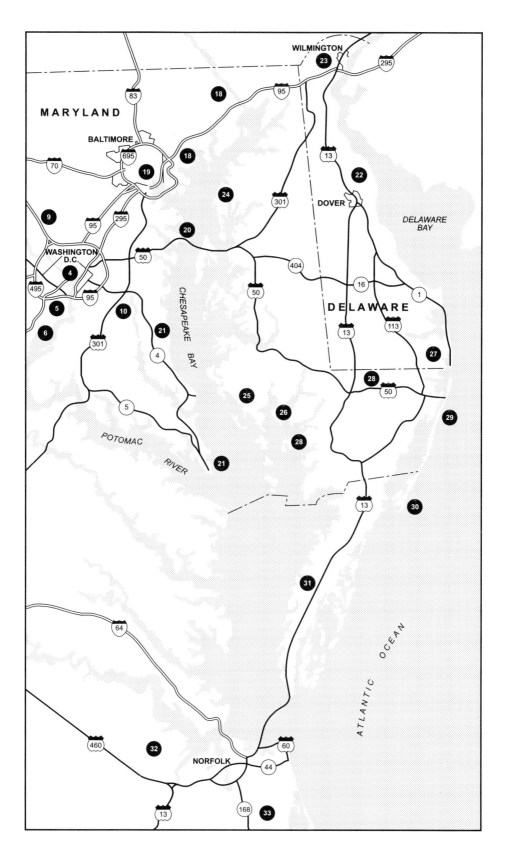

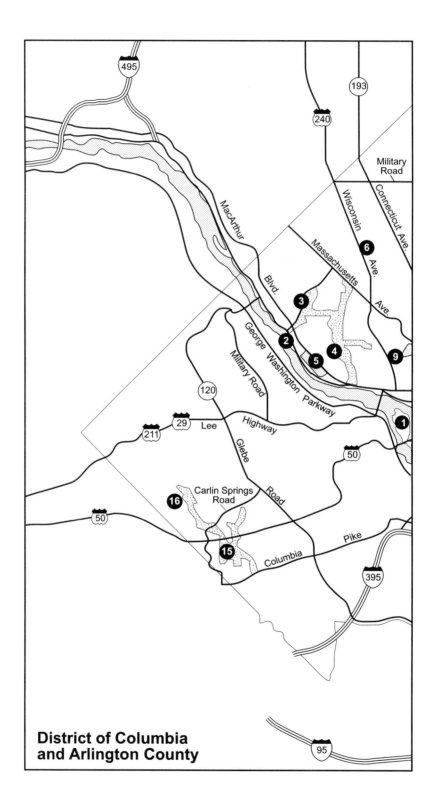

**District of Columbia
and Arlington County**

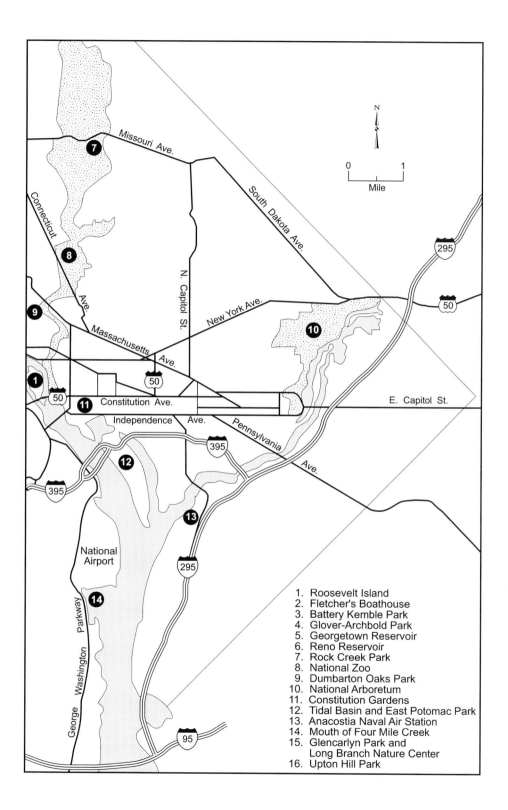

1. Roosevelt Island
2. Fletcher's Boathouse
3. Battery Kemble Park
4. Glover-Archbold Park
5. Georgetown Reservoir
6. Reno Reservoir
7. Rock Creek Park
8. National Zoo
9. Dumbarton Oaks Park
10. National Arboretum
11. Constitution Gardens
12. Tidal Basin and East Potomac Park
13. Anacostia Naval Air Station
14. Mouth of Four Mile Creek
15. Glencarlyn Park and
    Long Branch Nature Center
16. Upton Hill Park